Raised *by* Animals

Also by Jennifer L. Verdolin

Wild Connection

Raised *by* Animals

The Surprising New Science of
Animal Family Dynamics

JENNIFER L. VERDOLIN
FOREWORD BY MARC BEKOFF

THE EXPERIMENT

New York

The Experiment, LLC
220 East 23rd Street, Suite 301, New York, NY 10010-4674
theexperimentpublishing.com

This book is sold with the understanding that neither the author nor the publisher is engaged in rendering professional parenting advice to individual readers. It is not intended as a substitute for consultation with a health professional. Names and identifying details have been changed to protect the privacy of individuals.

Many of the designations used by manufacturers and sellers to distinguish their products are claimed as trademarks. Where those designations appear in this book and The Experiment was aware of a trademark claim, the designations have been capitalized.

The Experiment's books are available at special discounts when purchased in bulk for premiums and sales promotions as well as for fundraising or educational use. For details, contact us at info@theexperimentpublishing.com.

Library of Congress Cataloging-in-Publication Data

Names: Verdolin, Jennifer L., author.
Title: Raised by animals : the surprising new science of animal family dynamics /
   Jennifer L. Verdolin ; foreword by Marc Bekoff.
Description: New York : The Experiment, 2017.
Identifiers: LCCN 2016046676 | ISBN 9781615193448 (pbk.)
Subjects: LCSH: Parental behavior in animals. | Parenting. | Reproduction.
Classification: LCC QL762 .V47 2017 | DDC 591.56/3--dc23
LC record available at https://lccn.loc.gov/2016046676

ISBN 978-1-61519-344-8
Ebook ISBN 978-1-61519-345-5

Cover and text design by Sarah Schneider
Cover photograph © Suzi Eszterhas | Nature Picture Library
Wild Animal Engravings © Graphic Goods | Creative Market
Leaves © Vector Hut | Creative Market

Manufactured in the United States of America
Distributed by Workman Publishing Company, Inc.
Distributed simultaneously in Canada by Thomas Allen and Son Ltd.

First printing May 2017
10 9 8 7 6 5 4 3 2 1

To Ramie,
thank you for being my family

CONTENTS

FOREWORD BY
MARC BEKOFF

Nonhuman animals, we know now, are not so different from us. They grieve, they empathize, they get angry, and they experience joy. They also make decisions based on a sense of fairness and of right and wrong. I have made it my life's work to investigate the complexity of animals' inner lives—rich with intelligence, social fluency, and emotional depth—and to raise our consciousness so that we will change the way we view, appreciate, understand, and treat them. And, as cutting-edge comparative cognitive and behavioral research continues to reveal new ways we've long underestimated them, the imperative only grows for all of us to treat other animals with the empathy and compassion they truly deserve.

The penetrating and evocative science in the pages that follow shows us how far we've come in our understanding of animals. It also shows beyond doubt that a more sympathetic view of other species—from rats to birds, cheetahs to prairie dogs—is just as beneficial for us as it is for other animals. We have much to learn from them.

Consider, for example, the way animal parents teach their children the value of sharing. Many human parents wonder whether sharing can be taught, or whether they should let nature run its course and leave their children to decide for themselves how to cooperate with their peers. But if such parents were to turn their attention to the way parenting is actually practiced in nature, they'd see that the "natural" way is anything but hands-off. Galápagos fur seal mothers, for example, are quick to intervene to encourage cooperative behavior among their pups. When an older brother or sister pesters or attacks a younger sibling, mothers passionately defend their youngest and actively discourage fighting. "Natural" parenting doesn't imply cold-hearted instinct. Rather, it's based on principles of empathy and cooperation that we would do well to emulate.

Galápagos fur seals only scratch the surface; as *Raised by Animals* makes clear, there are many lessons to be learned from animals, if we're willing to overcome our biases against them. Furthermore, cultivating a greater sense of understanding and acceptance of animal parenting also encourages us to cast off the damaging biases we hold against members of our own species. Anyone who doubts the ability of same-sex partners to raise healthy children would do well to study Laysan albatrosses, many of whom very successfully grow up with two mothers.

Closely investigating animal behavior also reveals how our attempts at trying to preserve a sense of human exceptionalism can be not only groundlessly arrogant, but counterproductive. As you will read in greater detail, anthropologist and primatologist Dr. Sarah Hrdy points to evidence that, while nearly half of all primate species, not to mention most human cultures throughout history, have engaged in some form of cooperative parental care, contemporary Western society continues to hold dear the notion of a "traditional" family unit, with one mom and one dad raising their children by themselves. Once again, we'd be wise to look to the animals around us for inspiration as to how to be more supportive, and not so unduly taxed and stressed.

Accepting that other animals are multifaceted, thinking, feeling, reasoning beings is better for them, and it's also better for us. It's time we all learned to take a much closer look at how other animals navigate the complexities of parenting—I promise you'll soon see them, as I do, as fascinating individuals deserving of our love, respect, and attention.

MARC BEKOFF
Boulder, Colorado
December 2016

MARC BEKOFF, PhD, is professor emeritus of ecology and evolutionary biology at the University of Colorado, Boulder, and cofounder with Jane Goodall of Ethologists for the Ethical Treatment of Animals. He has won many awards for his scientific research, including the Exemplar Award from the Animal Behavior Society and a Guggenheim Fellowship. Marc has published more than a thousand essays and thirty books, and has edited three encyclopedias.

CHAPTER 1

It's All Relative

When I think of parenting, I think of Oma, my grandmother. When I would come home from school, Oma was there. When I got banged up playing tag football, fell out of the tree, or held out muddy hands to share my latest slimy creature, Oma was there. No matter what I was into or up to, Oma loved me warmly and unconditionally. She taught me gin rummy, smoked like a chimney, and drank coffee in lieu of water. Like a benevolent dragon, she would blow smoke out of her nose to entertain me. We were two peas in a pod. And still now, so many years later, thin German pancakes slathered with warm butter and sprinkled with cinnamon sugar arouse a yearning for her. That and a Dixie cup full of M&M's.

Why M&M's? Every day—every single day—she would dispense a small portion of M&M's into a cup for me after school. Of course, this meager portion was never enough. I knew where she kept them hidden (in the top drawer of her dresser) and I would sneak into her room and refill my cup. I was convinced that I was clever and sneaky, but now I'm pretty sure Oma knew what I was up to. For some, the smell of cookies baking, or other everyday moments, encapsulate childhood and evoke memories of home. They say home is where the heart is, and Oma was my home.

Although I also lived with my biological parent, Oma was my true parent until I was nine, when, for reasons I wouldn't understand for years to come, she moved back to Brazil. Her unexpected departure left me traumatized. I shared a total lack of affection with my mother and her new husband, let alone my absent father. The comparison occurs to me now, all these years and a PhD in animal behavior later, that I proceeded to act as any Hawaiian monk seal pup would when it gets separated from its mother: I tried to get adopted by another family. Before continuing on about monk seals, I should explain something about myself. My study of animal behavior, coupled with my tendency to think about the places where human and animal behavior intersect, has made it second nature for me to see similarities, construct analogies, and make comparisons between us and other species. I don't anthropomorphize other animals; I zoomorphize people, including myself!

Thus, it is easy for me now to think about my behavior as a child as though I were a lost monk seal looking for a new mother, something that happens frequently among Hawaiian monk seals. Sadly, they are perilously endangered with only about a thousand left. They are the only seal endemic, or native, to Hawaii. Like other seals, the babies are born primarily on beaches. However, unlike elephant seals or fur seals, monk seal moms aren't, shall we say, experts at recognizing their own pups. This could be disastrous, but for the fact that females willingly take care of *any* begging pup. It is possible that the females are in a hormonal state that stimulates them to provide parental care to offspring that are not their own. As you will see, in many other animals, recognition of one's own offspring is highly developed, allowing parents to be more selective. But for baby monk seals at least, in the chaos and confusion of a beach filled with hundreds of monk seal pups, if you become separated from your mom, not to worry. Cry loud enough and someone will take you in.

So, I cried, and my friend Stephanie's mom, Mrs. B, the quintessential Italian mother, short in stature but brimming with love, took me in. Mrs. B's house was *that* house. You know, the one where all the neighborhood kids converged. Where there was always food, a pool, television, seemingly never-ending sleepovers, and Mrs. B looking after everyone's children, not just her own. I say "Mrs. B's house," but there was a Mr. B. He was the kindly father figure, encouraging us to play pranks on the wife he loved so much. At Stephanie's house I was Juniper, not Jennifer. I was loved and cared for and valued for who I was. They say it takes a village to raise a child. Well, Mrs. B's house *was* the village.

Even at such a young age, I could recognize the striking contrast between my household and Stephanie's, and I was puzzled by the differences. Why was my own home so hostile? Why was it so unloving?

These unanswered questions would continue to preoccupy me well into my adult life. It wasn't until I was deep into my graduate studies that I began to understand the sources of tension between parents and offspring, of competition among offspring, and of the conflicts that are ever-present in many families, including my own. My dissertation research focused on the social behavior of Gunnison's prairie dogs. One late June afternoon, as I sat watching the newest generation of prairie dogs interacting with their parents, I witnessed a mother prairie dog stand upright on her burrow with her front paws draped over the narrow shoulders of one of her pups. It seemed like such a tender moment, but then, as the pup turned to nurse, the mom scampered off, pup still attached, bouncing upside-down along the ground until it could no longer hang on. I laughed out loud, thinking what a clever strategy this mom employed to communicate to her pup, "You're weaned!" It is moments like these that, beyond the sheer joy and fascination that comes with such observations, reveal to me the shared relationship we have with other species. The human condition is uniquely human—after all, a human mom might very well tire of nursing her child and extract herself in a way different from a prairie dog mom—but the underlying similarity remains.

I believe I was drawn to thinking about animal family dynamics because I wanted to comprehend my own family and upbringing. I learned that occasionally vervet monkey mothers reject their offspring, particularly if they are stressed or in poor physical condition themselves,[1] and that competition between siblings is more common than cooperation. I discovered that in a plethora of fish, birds, and mammals, fathers are active, involved parents, and sometimes do all the heavy lifting when it comes to raising the kids. I read about the family life of pilot whales, where sons and daughters live with their mother and extended family for life, whereas in the case of gorillas, both sons and daughters leave home to make their mark in the world. It was eye-opening to see these behaviors reflected in the animal world.

I admit, in some sense I was relieved to learn that parent-offspring conflict and sibling rivalry are just as central to animal parenting and families as protection and devotion. The comparison wasn't just entertaining; in many ways, it provided the tools to advance my own thinking. New questions took the place of those older ones. Instead of wondering why my family was so different from Stephanie's, I was now trying to understand why we, as humans, have such variety in the way we parent. I wanted to explore what conditions favor certain approaches and whether we are parenting in ways that create the greatest opportunity for success for our children. Animals do this all the time, and it wasn't clear to me that we humans are doing the same. I began to devote my research to these parallels between humans and animals in the realms of parenting and families, and to contemplate this area of our lives from the perspective of evolutionary biology.

For instance, when thinking about having children, how many of us actually consider the tension we will inevitably introduce between our needs and the needs of our offspring? That is, until we have them. And even then, how many parents are stopping to appreciate the finer nuances of the underlying biological and evolutionary origins driving their experience? As human beings we have an enormous capacity for self-reflection and cultural learning, yet few of us harness the power of comprehending the biological basis of behavior, which can transform our experience as parents. Instead of feeling guilt over the inevitable conflict we experience as parents between meeting our needs and those of our children, we can instead recognize the universality of such conflicts. By shifting the focus away from how we are supposed to feel (or not feel!) we can discover that there are myriad ways to cope.

Take my friend Julie, for example. I'd met Julie a few years earlier at a writer's picnic. We bonded instantly over our mutual affection for prairie dogs (not as surprising as it might seem since she had lived in Colorado for many years) and passion for writing. Originally from Scotland, she had fallen madly in love with an exchange student and followed him to America, where they married. When I met her, she had desperately wanted a baby, so when Sam finally came into this world, she was ecstatic. I hadn't seen her since he was born, and he was already two months old by the time we were finally able to schedule a lunch. We decided to meet at a café near her house for what she said would have to be "a very quick lunch."

She hobbled in, struggling to carry a cumbersome infant car seat and what seemed like an enormously large baby bag for such a tiny baby, and apologized for being late. Looking disheveled, she approached the table smiling, only to freeze in a sudden expression of horror as she blurted out, "Oh no, I must have left my purse in the car! Can I leave Sam here for a moment with you?" Before I could even nod my head, she was dashing off again. Such a dramatic entrance and exit, over a purse.

I looked down at the tiny thing in the car carrier. He looked angelic, sleeping peacefully in a onesie that seemed to be almost swallowing him up. How could it be that this completely innocuous-looking thing could be wreaking such total havoc on the perfectly put-together, punctual Julie I had always known? Then she was back, sitting down with a harried look on her face, saying, "I'm sooo sorry." Even though

I said it was no problem at all, she rushed to explain, "It was another one of his perfectly timed poops, I don't know how he does it. I was just clicking him into the carrier and *bam*. Diarrhea all over . . ."

I looked around uncomfortably. *Can you talk about diarrhea in restaurants?* I wondered. Okay, you probably *can*, but should you? Even if it's newborn diarrhea? But nothing was stopping her. "Right through his clothes, Jennifer, into the blankets. Is that normal?" I don't think she really wanted an answer, and truthfully I had no idea, but was I starting to think I really didn't ever want to find out. All I kept thinking was, *Why are so many new parents obsessed with poop?*

Her one-sided, half-crazed description of eight weeks of motherhood went on, and on, and on. "And look at me, Jennifer, I look like I've been eight rounds with a heavyweight champion—you just have no idea, no one does. I haven't slept for more than forty-five minutes at a time, my nipples hurt *all* the time, and I haven't even been able to find the time or energy to cut my toenails since I was in my eighth month of pregnancy!"

And this was just the first eight weeks. The constraints and costs of parenting may change over time, but they remain high. And yet parents often feel guilty about their occasional resentment over all their sacrifices. How to explain the psychology behind these conflicts?

Sigmund Freud framed parent-offspring conflict in a strange sexual fantasy sort of way, but in my opinion evolutionary biologist Robert Trivers takes a more sensible approach.[2] From an evolutionary perspective, it is advantageous to act in ways that benefit you—in other words, in ways that help pass your genes on to the next generation. That, of course, is where making babies comes in. The problem, which Trivers points out, is that your offspring are 100 percent genetically related to themselves and only 50 percent genetically related to you. This disparity is ultimately the source of conflict between parents and offspring. Why? Because parents, Trivers argues, balance their investment in any one offspring against their own survival and ability to produce future offspring. However, because each child is more closely related to itself than to its parents or potential siblings, they are focused on its own needs and survival. Therefore it may want *more* (food, attention, etc.) than the parents are willing or able to provide. This tension can become more pronounced if there are multiple offspring.

These days we don't incorporate children into our lives, but rather work the household around our children. This creates an even greater sense of conflict between parental needs and those of our children. It is tension of our own making, and it reflects the drive we have to help our offspring thrive. But many parents are *desperate* for their kids to succeed, and desperation creates anxiety, which makes us do strange things that far exceed a healthy approach to investing in our offspring.

Here, again, animals are instructive. They, too, invest heavily in their offspring, sometimes compromising their own health, as penguins are apt to do, although even they have limits. In general, though, many other species seem to have healthier boundaries when it comes to how much is too much.

On the flip side are parents who swing too far in the other direction, investing little more in their children than the bare minimum, and focusing instead on their adult needs. However, what we define as "adult needs" is very different in the case of humans than in other animals. A female polar bear loses over 40 percent of her body mass after fasting for eight months to give birth and sustain her cubs until they finally emerge in the spring. Starvation and poor physical condition may lead her to abandon the cubs after they are born to attend to her own needs to survive.[3]

In contrast, some of the "needs" that human parents are fulfilling when they neglect their parenting duties are not closely tied to authentic evolved needs. Say Dad is cheating on Mom, or Mom is cheating on Dad. Although unfortunate, it is "understandable neglect" from an evolutionary perspective, since this behavior can increase their chances of reproducing and diversifying their available gene pool. What about Mom or Dad plopping down in front of the television or

computer instead of interacting with their kids, or neglecting to prepare nutritious meals despite having the resources to do so? I'm not so
sure those are evolutionarily explainable, but they are likely a more
frequent source of inattention than marital infidelity.

Within families, parent-offspring conflict can show up in yet
another way: favoring one child over another. I know, I know, we don't
like to talk about it. And if one of our children remarks, "Mom, c'mon,
please tell me, who is your favorite?" the typical parent response is,
"You are each special in your own way. I couldn't possibly love one of
you more." But, sometimes parents play favorites.

This raises another advantage of looking at human behavior
through the lens of animals: It's a way to circumvent taboos. While
playing favorites is a well-studied phenomenon in animals, it's a largely
unacknowledged reality in humans. Take a look at the eastern bluebird. The males sing a sweet, tittering serenade for the females, but
that's where female favoring stops. Once the eggs hatch and the chicks
are fledging, eastern bluebird dads clearly favor their sons over their
daughters. When the fledglings venture from the nest, dads spend
more time protecting their sons from potential predators, leaving their
daughters exposed to danger. And, in the eyes of eastern bluebirds,
not all sons are created equal, either. Given two sons, a father bluebird will always preferentially protect the brighter-colored son.[4] Why?
Because it is advantageous to invest more in higher-quality offspring,
since they will have a higher probability of surviving and succeeding.

Long-standing studies show that human parental favoritism does
in fact exist, and in as many as two thirds of all families (we will discuss this in depth in Chapter 6). Psychologists, sociologists, parenting
experts, and bloggers can all tell us how favoritism is ethically wrong
and psychologically damaging, but what's missing is the answer to *why*
it is so common—not to mention real-world strategies to reduce its frequency in our own families. It might make some of us feel better to say
it happens because there is something iniquitous about the behavior
of those parents' parents—to trace it back to a previous generation.

But then does that mean that two thirds of all parents are inadequate or bad parents?

I'm going to have to side with Trivers on this one and suggest that perhaps there is something other than bad parenting going on here. Furthermore, unless we can grasp the underlying biological root of something so ubiquitous in both human and other animal behavior, we remain unable to implement effective counterstrategies—if necessary. Worse, we disown our behaviors, are plagued with guilt, and leave the wreckage for the next generation.

I propose, instead, that by looking at families through a biological lens we may uncover biological explanations for many of the dynamics that seem so contradictory to our *beliefs* about what families ought to be, how parents are supposed to feel, and what drives many of our own behaviors and attitudes as parents and as children. By looking through a biological lens, examining behavior from an evolutionary perspective, and using our animal counterparts to illustrate fundamental principles, we will see that there are many reasonable solutions to this thing called parenting. Some animals converge on similar solutions, while others have positively ingenious ways of handling some common parenting dilemmas. And I've found that there is tremendous variation in parenting styles not only across species but also *within* species. Yes, there are bad parents, good parents, and even better parents, but the devil is in the details. And that is what this book is about: what it means to be a parent, no matter the species, and the ways parents build relationships with their children to create their own versions of a family.

As any expectant parent or new parent knows, there is a dizzying whirlwind of endless information out there about how to parent or not to parent: a frenzy of books, near-infinite expecting-mommy blogs, and modern-dad blogs with such titles as "The 10 Things You Need to Know," "The 12 Things You Should Never Do," "The 7 Things You Should Already Be Doing," and "The 100 Things You're Most Likely to Forget." Help!

Raised by Animals is not another parenting book touting the latest "secret"—to the contrary, it rejects the premise that there is such a thing. Instead, it takes a closer look at what it means to be human, to be parents and children. We're heading back to nature, our nature. There are vastly different human cultures on this wonderful planet, which we humans share with an endless number of other species. By highlighting the variation that exists within our own species and comparing that with what we see across the animal kingdom, perhaps we can expand and embrace a broader view of family and find the threads that tie us all together.

I am not attempting to equate the complex and intricate nature of being a human parent to that of, say, a sea turtle parent, who on the surface appears to care little about the fate of her babies. After all, she digs a hole, lays the eggs, covers the hole, and departs, never to know the fate of her children. But even the seeming simplicity of this approach to parenting is deceptive, because the female sea turtle that comes upon a beach to lay her eggs has faced almost insurmountable odds just to make it to adulthood and reproduce. If she survives long enough to have mated, she must then emerge from the water, bringing her enormous body, buoyant in the water but cumbersome on land, ashore. In doing so, she places herself in grave danger. From there, she must decide where to lay her eggs. She does not choose this place haphazardly. She carefully selects what she believes will be the optimal location to protect the developing hatchlings until they are ready to fight their way through the sand to the surface and scurry off in search of the ocean.

She digs awkwardly with her flippers to just the right depth—not too deep, not too shallow. With copious tears streaming from her eyes to eliminate the excess salt in her body, she lays 80 to 120 or more eggs. Once she deposits her precious cargo, she covers the eggs and lumbers off, back to sea. A single female may do this once, twice, up to as many as five times in a single season. Thus, direct parental care is just one form of investment, as the female sea turtle can attest to. And like any parent, some do a better job of it than others.

At the other extreme, male and female pilot whales stay with their mother for their entire lives, which can be upward of fifty years! The grief we witness when killer whales are separated from their families, when elephants frantically band together to rescue another's child, when a gorilla mother steadfastly refuses to lay down the body of her dead infant, and when an emperor penguin wails desperately in the hope that her cries will rouse the frozen chick lying at her feet—these things reveal more about what we have in common than where our differences lie, and they suggest that our emotional experience of parenting emerges from a deep connection we have with other species.

Even so, some say that comparing animal and human behaviors is like comparing apples to oranges. To that, I say: *Exactly!* Though apples and oranges diverged genetically over eighty million years ago, they are both from fruit trees, start off as a single flower that attracts insects to pollinate it, and grow with water and sunlight to have a similar circumference, diameter, and weight. They are both sweet (well, except maybe for Granny Smiths) and can be juiced. They both create seeds as a means of reproduction. And for the calorie conscious out there, they both have approximately 115 calories. In the proper context, such a comparison can reveal fresh insight. So, I'm all for comparing apples and oranges—it's all relative. More importantly, researchers, like myself, are constantly discovering new information about other species that are blurring the lines between ourselves and other species. Pigeons recognize words, dogs read our emotions and understand our words, horses communicate with us, and fish sing at dawn. These are just a few of the discoveries reported in a single month while writing this book. As these boundaries continue to collapse, it makes even more sense to look to other species for some counsel.

Ultimately, we share an evolutionary history with every single living creature, even sea turtles and pilot whales. There are ways in which we are similar and there are ways in which we are different.

We can leverage these relationships to reflect on different areas of our human lives. At the same time, we must be cautious of the "is-ought" trap, or Hume's Law, in which one makes an inference that because something exists, it is, or ought to be, a moral truth. When looking to biology, this is essentially the same as committing a naturalistic fallacy, inferring that because something is natural it is morally acceptable. This erroneous leap implies that what is found in nature is good, with "good" meaning whatever we, as a society, have decided is morally or ethically proper. It completely discounts the context.

To illustrate this point, we can look to coots and moorhens. Both are small waterbirds that hatch a large number of chicks at a time. The parents usually feed the chick that is closest to them, but if one chick becomes too "demanding" the parents may discourage the chick by picking it up and shaking it. Sometimes a chick may die.[5] From a Darwinian, natural-selection perspective, it is usually maladaptive (and therefore rare) for animal parents to kill their own offspring, yet occasionally it does happen. The naturalistic fallacy in this case would be to say that because coots punish their offspring in this way, it is evidence that corporal punishment is good and morally correct.

However, there is value in understanding where our behavior comes from and the ways in which we are the same or different from other species. By placing behaviors in an evolutionary context, we gain insight into why certain approaches may not make sense from a biological perspective. This, in turn, allows us to gain a different outlook and, in some cases, possibly make choices that produce more successful outcomes.

As I enthusiastically delved into the jungle of this thing called parenting, my research not only expanded my understanding of what it means to be a parent, it also forever altered how I conceive of my own experiences as a child. I uncovered some unexpected similarities and peculiar differences between humans and animals when it comes to being a parent, being a child, and being part of a family, as well as just some downright odd and spectacular approaches to parenting in other species. More than that, just as with dating and relationships, talking about pregnancy in male seahorses, sibling rivalry in sharks, and coot parents that punish greedy chicks provides the ideal transition to discussing some very sensitive parenting topics. Because, let's be honest, discussing parenting is probably the only topic that is fraught with more danger and sensitivity than talking about the right or wrong way to date and have a successful relationship.

We'll begin by taking some baby steps. First, whether we are discussing how some frog moms have their tadpoles develop in their stomach, only to regurgitate them out of their mouth later, or sharing the unusual solution that some legless amphibians have evolved to "nurse" their young, there is no shortage of the strange and fascinating when it comes to parenting—including among humans! Then we'll start to explore how the topics and examples covered may directly help to explain why particular animals parent in particular ways and shed light on our own behaviors, potentially offering alternatives to how we parent our children.

Parts of this book may shock you, parts will delight and entertain you, and parts will challenge many of the notions we have about raising kids and families. Along the way we will uncover many surprising unconscious biological factors driving our experiences, both as children and as parents, and learn how to apply these concepts to becoming better parents, ones saddled with less guilt and resentment. To do this, we'll look into all aspects of parenting, including how animal parents deal with discipline, how animal siblings negotiate their relationships, and the important role of fathers. This book will make you laugh, gasp, and help you discover, develop, and implement behaviors and strategies that enhance your ability to be an effective parent. In the end, I hope it helps parents to accomplish what the vast majority of us set out to do: raise happy, healthy, well-adjusted, successful children— while surviving and remaining sane in the process!

CHAPTER 2

We're Pregnant!

Growing Bellies, Nest Building, and Arrival

I'll never forget my first run-in with possible pregnancy. I was single, young, a mere eighteen years old, and frightened after an incident involving an epic birth control fail. A few weeks and ten ragged fingernails later, I was officially "late." And by late I mean late for someone who never paid much attention to the rhythms of her cycle. The stress, the worry, the what-ifs, even the excitement are often all that is required to keep your period at bay. Finally, with my best friend in tow, I was off to the drug store.

Fortunately for us, one can't be a little bit pregnant, and one of those precious pregnancy sticks would give me an answer. We bought a three-pack, just in case, and the all-too-familiar routine ensued. Pee, wait, look, repeat. Whether it is a double line, a plus, or some other marker, the tests are made to be easy to decipher. That time, and subsequent times I felt compelled to use a pregnancy test, my test results were negative. Like magic, my period would materialize within hours, as if it had been there waiting all along, teasing me. I doubt that other species agonize over finding out whether they are pregnant, whether they should be happy about it, and what they will do about it. They just are expecting.

That is not to say that the ways in which other species get pregnant, experience pregnancy, and deliver the next generation into the world is as straightforward as you might think. Getting pregnant varies in some fascinating and sometimes unexpected ways. Our road to pregnancy usually involves sex, hopefully good sex. That isn't always the case for other species. It definitely wasn't for the Australian gastric-brooding frog. I write that in the past tense because this remarkable amphibian has been listed as extinct in the wild since the 1980s. This creature looked unassuming enough, kind of a brownish orange. You might expect that a frog that regurgitated its fully developed tadpoles out of its mouth would have some spectacular distinguishing physical feature. But no, it looked like a somewhat ordinary frog. Clearly, female pregnancy in this species was anything but conventional.

First, female Australian gastric-brooding frogs didn't get to have sex. The female would deposit roughly forty eggs—whether in land or water, we are not sure—and a male would come along and plop his sperm all over the eggs to fertilize them. But the fun didn't stop there. Next, the female took each fertilized baby frog-to-be and, one by one, ingested them. Yes, you read that right. Mama swallowed her future tadpoles. This is where things really got interesting. If you think that the acids in the stomach would be lethal to her developing embryos, you would be correct. That is why the young, developing tadpoles excreted a hormone called prostaglandin E2, which suppressed the release of hydrochloric acid normally found in the mother's digestive system.[1] This continued for eight weeks until the tadpoles emerged, sticky with mucus, from their mother's mouth. Aside from the obvious discomfort these devoted moms must have experienced, this process also meant that mother's didn't get to eat for those eight weeks. I bet regular old human pregnancy and labor is starting to sound pretty good right now. Incidentally, this hormone, prostaglandin E2, is the same hormone used to induce labor in human females.

This is just one example of the extreme physiological demands pregnancy places on individuals. It also highlights a relationship between the developing offspring and their parent that is a delicate balance between cooperation and parasitism.

I suppose with all the variations on pregnancy out there, there is something to be said for us having escaped this particular form. That doesn't mean, though, that all pregnancies in humans are identical or easy. Talk to any two women who have had children and you will quickly discover that no two pregnancies are the same. Some women positively glow, as if they have a halo of light surrounding them, while others may spend the entire time emitting an altogether different hue: green. In thinking about this special connection we share with so many other species, perhaps mouth-brooding frogs excluded, it occurred to me that, unlike other animals, we can talk about our experiences about being pregnant and going into labor. We can compare notes, share stories, and try to figure out what's normal. I began to wonder how our experience fit into the grander scheme of things. How do other species experience pregnancy? Do they get morning sickness? Food cravings? Is labor hard for other animals? And finally, how can understanding these matters help us understand our own experience?

Pregnancy: Boy, Isn't This Fun?

In an ideal world, you find a terrific partner, your best match, and ride off into the sunset to make a baby. Making a baby is the easy part, right? Have sex and voila, you become pregnant. The reality, however, is that a tremendous number of things have to work in concert, under just the right conditions, for a female to become pregnant. There is timing sex to coincide with estrus, that magical time when a female is physiologically receptive to becoming pregnant. If we do not become pregnant, we menstruate, while most species simply reabsorb the endometrium and go through another cycle of estrus. The length of cycles varies from species to species. When it comes to ovulation, some species must be coaxed, whereas, for the most part, we spontaneously ovulate every month. A precious egg is released, ready and waiting for just the right sperm to find it. The egg, however, does not like to be kept waiting and only hangs around for about twenty-four hours. A narrow window of access, to say the least.

This is not terribly uncommon. Cape ground squirrels are also spontaneous ovulators. Despite their name, Cape ground squirrels' range extends well past Cape Town, South Africa, covering parts of Botswana and Namibia. They like it dry and are well adapted to life in the desert, getting most of their water from the foods they eat. Similar to my beloved prairie dogs, they build burrows and sit on their haunches in a posture we call "posting" to keep an eye out for predators. What is notable about them is how short a time period the female's egg is friendly to sperm: a mere four hours.[2]

Be it Cape ground squirrel or human sperm, and assuming it's a sperm-friendly time, the hurdles for these little swimmers are only just beginning. Having to swim the equivalent of the English Channel, each sperm faces almost insurmountable odds. Given how the structure, chemical composition, and immune response of the female reproductive tract is hostile territory for sperm, it's a wonder that anyone gets pregnant at all. There should be a sign that reads ENTER AT YOUR OWN RISK, warning sperm of their impending death. It's rather like *Indiana Jones and the Temple of Doom*—booby traps everywhere. As they make their way toward the cervix, the gatekeeper of new life, sperm have to traverse an acidic environment. Many, many of the millions of sperm contained in ejaculate fall by the wayside along the way. If they make it to the cervix, it doesn't get any better. They will encounter cervical mucus and an overwhelming "anti-sperm" immune response of leucocytes and antibodies rushing in to coat the sperm, shutting them down on the spot. The leucocytes gobble up the offending sperm.

There are a few hypotheses to explain why female bodies respond so forcefully to keep sperm out. One idea that has quite a bit of traction is that deformed sperm are blocked from entering the cervix.[3] This allows females to be very selective and prevent unfit sperm from getting to their egg. And it doesn't stop there. The egg, or ovum, changes its chemical properties to make it difficult for sperm to penetrate. Of the 100 to 150 sperm that complete this journey, only one can penetrate this coating and succeed in having its plasma membrane fuse with the egg to create the activated cell that will be your future baby. Consider for a moment how, of the millions of sperm released, there is a success rate of roughly 0.002 percent that ever set their sights on an egg in the first place. Whew, tough odds. Given all of that, it is a wonder we manage to conceive at all. So, if you find yourself having a hard time getting pregnant as fast as you might like, relax and breathe, because getting pregnant isn't as effortless as your middle school sex education class suggested.

For humans, pregnancy is undeniably linked to being female. Men may cringe at the thought of delivering a nine-pound baby, and frankly, who wouldn't, but so as not to have my male readers feel left out, there are plenty of animal males out there in charge of giving birth to new life—specifically, pipefish and seahorses. The big-bellied seahorse seems appropriately named and was in the news recently when researchers discovered that pregnancy in big-bellied males is more similar to that of human females than was previously realized. They are the largest of the seahorses and like to hang out in shallow Australian reefs, where they attach to seagrasses, algae, and even sponges.

Technically, males aren't pregnant in the truest sense of the word. But they are in charge of caring for the developing zygotes in a specialized pouch where fertilization also takes place and the baby seahorses implant on highly vascularized attachment sites. The most fascinating aspect of their pregnancy is that the brood pouch is composed of specialized tissue, much like the human uterus, and undergoes changes in shape and structure once the eggs are deposited inside. The male incubates the eggs, regulating the environment by controlling the pH, temperature, and oxygen levels. And here's the kicker: He provides nutrients, in the form of glucose and amino acids, to his developing children through the attachment sites. Aside from the superficial similarity, when scientists took a closer look, they also discovered some of the same genes were involved from conception to parturition in male seahorses and human females.[4]

Among human males there is a form of sympathy pregnancy. This is technically called couvade syndrome, and its existence is hotly debated. Strangely, there is the accusation that it is all psychosomatic, which, of course, some of it may be, but there are cases where males gain weight and have hormone fluctuations that mirror what is happening in their pregnant wives. I remember one boyfriend in particular who would always tell me, "I think you are about to get your period." He said this not because my disposition became cantankerous, but rather because he would gain about five pounds. When he magically lost the extra weight, it was an excellent predictor that I would begin menstruating within a day or two. This didn't happen once or twice; it happened every month. Luckily for him, he didn't get me pregnant—otherwise, who knows, he might have gone into sympathy labor for me!

Common marmoset dads are incredibly involved fathers, helping with delivery and carrying the little ones around. They're an unassuming New World primate weighing in at under a pound. Their long tails are all for show since they aren't prehensile, and they have to run like squirrels through the treetops. The large ear tufts make their heads appear larger than they really are and, along with other members of the Callithrix, they usually have twins! Although I will talk more in the next chapter about the changes that happen to both males and females to get ready to become parents, let's just say putting on some pregnancy weight isn't only for females and human dads. Gestation in this species lasts about five months. At the start of the female's pregnancy, males weigh a little less than a pound. Nothing really changes for the first month, or the second, or even the third. But by the fourth and fifth month, look out: Expectant dads are getting positively pudgy, doubling their weight right before the females give birth.[5] Understanding that males can be strongly influenced by the pregnancy of their partners may help some dads-to-be cope if they find themselves feeling off, more emotional than usual, or gaining weight even though they are hitting the gym or trails regularly. For obvious reasons we tend to pay more attention to what is happening to women during this time, but it is important to remember that men, too, are experiencing substantial changes.

Regardless of when they officially start, who is carrying the burden, or who is gaining the weight, some pregnancies can be surprisingly brief, while others last an awfully long time. In mammals, the winners of the shortest gestation period are members of the opossum family and the eastern quoll, native to Australia. The duration for these marsupials is a spectacularly brief twelve days. The yapok, or water opossum, is incredibly rare. Made for the water, it is found in the northern regions of South America and is the only semi-aquatic marsupial in the world. It is a peculiar creature, active mainly at night, feeding on fish and crustaceans along heavily tree-lined riverbanks. Being a marsupial lends itself to brief pregnancy, because like the eastern grey kangaroo, neonate water opossums have to climb the proverbial fur ladder into the pouch, where it is another two months before they get to ride on the outside on their mother's back.[6]

The longest pregnancy award for mammals goes to the African elephant. Imagine being pregnant for close to two years and then delivering a baby that weighs over two hundred pounds. Okay, I know relative to the size of elephant mothers, which can reach upward of nine thousand pounds, two hundred seems trivial. I wonder if, at some point over those two years, a female elephant ever wearies of being pregnant for such a long time.

I know that my friend Adrian did. Given my profession, many of my friends are also scientists, and Adrian is no exception. She is an anthropologist specializing in primates, focusing on, of all things, mothering! And yet, despite all of her training, we can take comfort in the fact that even she eventually felt what so many mothers do: *Enough already! Let's get this baby out of here!* At around eight months she felt as though she had been pregnant forever. As she so eloquently described it, initially it was like an alien growing inside her body, and by the time she was close to the end it seemed as though she had a squirrel under her shirt that couldn't get out. Squirrels are clever and charismatic, but no one wants one trapped under their shirt. Even though Adrian was, and is, ecstatic with the outcome—a beautiful baby girl, now over a year old—she did not enjoy being pregnant one bit, especially not in those late stages when she'd had enough. As a bonus, her misery was extended because her daughter arrived one week "late."

We have this strange need to predict everything, including the exact day our new baby will arrive. Most mothers seem to treat the date they are given as an estimate, until they are really done with being pregnant. Then, like Adrian, they view the date as the target and experience anxiety, despair, frustration, and disappointment when their baby doesn't arrive "on time." A doctor usually gives a woman a date of 280 days past her last period. Let's see what the average is.

The average gestation length in humans is 267 days, or roughly nine months, with an enormous spread of +/- 37 days. (By the way, this range does not include errors in estimating the time of implantation.) As with virtually everything else, there is variation from one individual to the next. And five weeks is a heck of a lot of variation! What is the source of this variation? Some babies take longer to attach to the uterine wall and those babies take longer overall to come out into the world. In older mothers who carry a baby to term, it tends to take longer to get to the delivery date, and if a woman was a big baby herself, she is more likely to have a pregnancy that lasts longer. [7] Mothers-to-be would be wise not to get too attached to a particular due date.

If the length of an elephant's pregnancy isn't enough to make you feel better about being human, perhaps one more example will help: the frilled shark. These sharks are not the most attractive creatures, having a face that, well, only a mother could love. Their jaws, complete with rows and rows of needled teeth, are positioned on the end of their nose, which has two large vertical slits. They have really large eyes, most likely because they inhabit the deep sea and there isn't a lot of light down there. Which, let's face it, in their case might be a blessing. Things don't get any better when it comes to their body shape, which is like an eel with rounded fins attached at various points along the back end of their body. This shark gets its name from the six frilly gill slits around its throat that look like *bouillonné,* the fancy ruffling or ruching of fabric common to Victorian era clothing.

As if all of this weren't enough, female frilled sharks are pregnant for over three years! The official term we use to describe mammals who give birth to live young is *viviparous.* These primitive living fossils do give birth to live young, but they are not placental like us. Instead they are *oviviparous,* which means baby frilled sharks develop inside an egg, like a chicken, but are kept inside the uterus until they are ready to hatch. You could watch paint dry faster than baby frilled sharks develop. They grow at an extraordinarily slow rate of half an inch per month and don't emerge until they are one-and-a-half to two feet long.[8]

Although I am not sure any woman wants to be pregnant for three years straight, some women, unlike Adrian, love to be pregnant. That was certainly true of Amanda. I met Amanda on a flight from Raleigh to Denver. She had four children, the youngest was five, and she confessed she missed being pregnant and wanted to have another child so she could experience it all again. Her husband, on the other hand, was not so keen to add to the brood. Sometimes when I am talking to someone, I begin to think of an animal that they are reminding me of. It happens all the time, actually. The wistful look on Amanda's face as she chatted with me about the joys of being pregnant made me think she wanted to be an eastern grey kangaroo.

Eastern grey kangaroos are marsupials found in Australia and Tasmania. For most of us, jumping or hopping requires a tremendous amount of effort. Try it now and you will see what I mean. Kangaroos have an interesting feature in that the tendons in their legs act like springs that allow them to achieve high speeds with minimal effort. How fast? Over 35 mph!

They don't start out in life this powerful, though. The miniscule baby kangaroo, or joey, emerges weighing approximately 0.03 ounces and is smaller than a jelly bean. Despite its diminutive size, this underdeveloped newborn demonstrates herculean strength, pulling itself up what is proportionally the distance of a football field to reach its mother's pouch.

It takes well over a year for one joey to become fully independent, but this doesn't hold mother eastern grey kangaroos back; having three vaginas helps kangaroo moms stay ahead of the game. Along with wombats and Tasmanian devils, kangaroos have a tripartite vaginal canal: Two are dedicated to the sperm delivered by a two-pronged penis. As if that isn't strange enough, a fertilized egg is sent to one of her *two* uteri. This allows the mothers to have a second baby before the older one is even out of the pouch. But this also means both joeys have to share the space.

And kangaroo mothers aren't the only ones on pregnancy double duty. Northern elephant seals have a nifty little trick called delayed implantation, which is the equivalent of being a little bit pregnant. They can give birth to one pup, then mate, and then keep the embryo in suspended animation without implanting into the uterine wall. There are several advantages to holding off on gestation. In the case of the northern elephant seal, and other seals, this delay ensures that the next pup will be born at just the right time. Too bad we can't schedule full-blown pregnancy, though there is some debate about whether or not it has happened in humans and may explain longer-than-normal pregnancies.

An inhospitable uterus could potentially lead to delayed implantation in humans, and a recent study suggests that it may be a trait common to all mammals.[9] For modern humans, psychological stress may be the most common cause of this phenomenon by elevating stress hormones in the body, which in turn decrease the levels of hormones necessary to make the uterus friendly toward an embryo. There has been a clear-cut case of delayed implantation in humans, but unlike the benefits provided in other species, it seemed to compromise the viability of the pregnancy.[10] Once again, if you are finding it difficult to get pregnant and infertility issues are not present, examining the degree of stress you are experiencing and reducing it may go a long way toward initiating and supporting a healthy pregnancy.

An interesting question, often raised by many moms-to-be, relates to how active they should be while pregnant, barring issues such as pregnancy-induced high blood pressure (preeclampsia) or other conditions that require bed rest. Some soon-to-be-parents don't do much while they are pregnant. There is the classic joke that pregnant women all lie around all day and eat pickles and ice cream. More about cravings in a little bit, but to be clear, even before the contemporary cultural revolution reintroduced women to the workforce, the vast majority of pregnant women around the world were hardly idle. The same cannot be said for most vipers. The majority of these snakes stop eating and limit their activity. There are a few reasons scientists believe inactivity is par for the course for many snakes. First, snakes must thermoregulate to digest their food, and their ability to do so effectively while pregnant may be compromised. Another possibility is that there simply isn't enough room in the female's body to accommodate her babies *and* food.

Not so for the western diamondback rattlesnake, though. A true working-while-pregnant mom, this species hustles for a meal throughout pregnancy. Rattlesnakes are pit vipers, owing to the pair of heat-sensing pits they have between the eye and nostril on either side of their head. All pit vipers are venomous. I was privileged enough to see my very first rattlesnake in June 2015. I was in Fort Collins, Colorado, a hip, cool city, and was working on a documentary filming prairie dogs. I knew there were rattlesnakes in the area, but one afternoon the cameraman and I were so engrossed in a discussion that we nearly missed seeing the five-foot rattlesnake crossing the path in front of us!

At any rate, like many snakes, rattlesnakes are viviparous, which you may recall means giving birth to live young, and as in the case of the frilled shark, the eggs are contained in the body cavity, where baby rattlesnakes rely on the yolk for nutrition while they develop. How do you tell if a female rattlesnake is pregnant and whether or not she has eaten? By radio-tracking them, naturally. As you can imagine, rattlesnakes are often slithering around every which way. By tracking them, scientists can find them regularly enough to see whether or not individuals have a characteristic bulge that appears after they eat and determine if they've had a litter of babies by finding them within the time frame that babies are expected.

By following snakes around, one study revealed that in a single decade twenty-seven females only produced forty-eight batches of babies. Total. That means, on average, female rattlesnakes reproduce once every five years! And they don't make a pile of babies either. On average they each produce four or five babies.[11] This should give people who support mass killing of rattlesnakes some pause. We need snakes—they are vital to a healthy ecosystem—and at the rate they reproduce, they need all the help they can get to escape becoming endangered.

To make these four or five babies twice a decade, pregnant female rattlesnakes, in contrast to other snakes, actually ate *more* often, implying a huge metabolic cost to being pregnant. So next time a pregnant woman claims she is eating for two, remember, just like a rattlesnake, it's sort of true. However, being pregnant isn't a license to eat for two full-grown adults! In general, as a species, we are in the midst of an obesity epidemic. When it comes to pregnancy, obesity is a real problem. Not only does it make it more difficult to become pregnant, but also women who are overweight experience more complications during pregnancy.

To gain a little perspective, it is helpful to talk about giraffes. Giraffes are darling and remarkable, and not just for their height. They embody the African savanna, though currently they are at risk of becoming endangered and roaming the open plains and woodlands no more. On average female giraffes stand eleven feet and weigh 1800 pounds, with a gestation period of approximately fourteen months. Now, newborn giraffes have to be somewhat self-sufficient shortly after birth because they need to stand to nurse and follow mom around. To this end, when they are born they are roughly six feet tall and weigh in at 200 to 250 pounds, which equals 11 to 14 percent of the mother's weight. A pregnant adult female giraffe doesn't gain much extra weight, and often in zoos the first clue that she is even pregnant is when she goes into labor. But let's tack on an extra fifty pounds as padding for our thought experiment, bringing the percentage weight gain to a max of 17 percent of the mother's starting body mass.

In humans, the current medical guidelines from the National Academy of Medicine (NAM) are as follows:

- Underweight women (body mass index [BMI] less than 19.8) gain 35 to 45 lbs.
- Normal-weight women (BMI 19.8 –26.0) gain 25 to 35 lbs.
- Overweight women (BMI 26.1–29.0) gain 15 to 25 lbs.
- Obese women (BMI 29.1–39.0) gain less than 15 lbs.
- Morbidly obese women (BMI higher than 39.0) gain no weight.

The average neonatal birth weight in the United States has declined slightly—a 2013 study shows that it was 7.58 pounds in 2000 and 7.47 pounds by 2008.[12] If we do a little math using the average height of women in the United States as 5 feet 4 inches, with a recommended weight of 110 to 144 pounds (depending on BMI), we can see that even at a starting weight of 144 pounds and the upper end of rec-ommended weight gain during pregnancy, human females would be gaining 24 percent of their starting body mass during pregnancy. A bit more compared to giraffes, but that doesn't sound too excessive. The question is, is this what we are doing?

No. For instance, in one study, normal-weight women gained anywhere from 35 to 50 pounds, a range of 24 to 35 percent using our imaginary numbers above.[13] A 2015 report by the Centers for Disease Control and Prevention (CDC) indicates that almost half of US women gain excess weight during pregnancy, though other studies report numbers as high as two-thirds.[14] Why is this a problem? Several reasons. Not only does it lead to complications for the mother during pregnancy, as already mentioned, but there is substantial evidence that excess weight before and excess weight gain during pregnancy has a negative effect on a child's health, increasing the probability of obesity, which can extend into adulthood.[15] As if poor health and a greater likelihood of disease were not enough, women who are obese during pregnancy give birth to babies that are considered biologically older by as much as a decade. By not recognizing the seriousness of maternal effects, we are crippling our children before they are even born.

WILD LESSONS

* Getting pregnant is not effortless. Conditions inside and out have to be right. If getting pregnant is taking longer than expected, check with your doctor to rule out infertility, get to know your cycle, and most importantly—relax!
* Men can experience pregnancy symptoms right alongside their partners. Better educating them to recognize and cope with symptoms helps *everyone*!
* When your doctor gives you a "due date," remember that this is an estimate with a potential thirty-seven-day spread. And when you start feeling like you have been pregnant forever with a twitchy squirrel trapped under your shirt, remember that at least you aren't a frilled shark.

continues . . .

- Pregnancy comes with enormous energy demands, but that isn't a free license to gain too much weight. Other animals gain little extra weight beyond the size of their newborn.
- During pregnancy, unhealthy weight gain causes long-term health consequences for your baby.
- Get moving and receive nutritional counseling during your pregnancy. However, if you need to lose weight, do it *before* you get pregnant.

If giraffes gained 35 percent or more of their body mass during pregnancy, well, there probably wouldn't be any more giraffes since they wouldn't be able to run fast enough away from lions.

When Coffee Smells Like Sh*t and Other Food Oddities

Speaking of food and eating during pregnancy, along the way some interesting things can happen. One of the most common is "morning sickness," though in reality it can occur unbridled throughout the day. In the medical literature it goes by the acronym NVP, which stands for nausea and vomiting during pregnancy. About 70 percent of women go through this, primarily during the first trimester—which is when all the major organs are forming—but some women experience it throughout their pregnancy. Rarely, things can get really bad and some women will develop extreme NVP, called hyperemesis gravidarum. How susceptible you are to NVP, how long it lasts, and whether it progresses to the more serious form is highly heritable. In fact, for NVP, something that almost never happens in science was reported: an almost perfect correlation between severity, duration, and genetics.[16] If you are one of those poor women suffering as you read these words, it's your mother's fault.

As with most things evaluated from an evolutionary perspective, we must ask if the benefits of feeling sick and vomiting in early pregnancy outweigh the costs. If you end up on the plus side of the equation, then the behavior or trait might be considered adaptive. If you're on the negative side, well then, the matter in question may have been beneficial at some point, but has now become maladaptive.

We might be inclined to assert that of course it is abnormal for a little over two-thirds of women to be ill, unable to eat, and throw up regularly at precisely the same time they need nutrients to feed the demands of a developing fetus. But is there a way for this to make sense? Margie Profet, a physicist by training, thought so and proposed that morning sickness wasn't an illness at all, but instead was an adaptation that protected embryos from disease.[17] The formal name for this hypothesis is "maternal and embryonic protection."

One way to evaluate whether nausea and vomiting during pregnancy may serve a particular purpose is to look for patterns or specific foods that trigger this unpleasant response. The most frequent foods that trigger disgust and nausea include meat and strong-tasting vegetables. To make sense of why these particular foods are involved, we need to go back in time a bit to when there was no such thing called refrigerators or genetic engineering. Though we may not think of it every time we take a bite out of another animal, their meat has enormous potential to carry pathogens and parasites and to cause illness. Here are just a few: *E. coli, Salmonella, Listeria* (very dangerous to pregnant women), *Campylobacteriosis, Shigella, Staphylococcus aureus, Giardia duodenalis, Cryptosporidium parvum, Toxoplasma gondii, Trichinella spiralis, Taenia saginata,* and *Taenia solium* (tapeworms). Yum.

What many people may not realize is that pregnant women are in a state of modified immunity. This is not a system-wide shutdown of a woman's immune response, as was previously believed, but rather the immune system is accommodating the developing fetus. Why would there be any involvement of a woman's immune system when she becomes pregnant? The idea initially put forth by scientists was that the developing baby is a bit like a foreign object, at least half foreign. All babies conceived the old-fashioned way share only 50 percent of their genes with their mother. The other 50 percent belong to a stranger, in terms of the mother's immune system. It was thought that a woman's immune system went on hiatus so as not to reject this strange creature.

We now have evidence to suggest that some highly specific communication is happening between the fetus and its mother through the lifeline, the placenta. Contrary to earlier hypotheses, it is now proposed that the immune system is not suppressed, but is tightly regulated.[18] Therefore, with the exception of a few bugs (e.g., listeria and malaria), pregnant women are not *more* likely to get an infection. Having said that, it is pretty well established that if a pregnant woman does become sick, either with the flu, other virus, or a food-borne illness, the consequences are more severe.[19] Herein lies the hidden cost. There are necessary tradeoffs in a mother's immune system to keep her body from rejecting the developing baby, some of which compromise a woman's ability to fight an infection when she is pregnant. Given that meat was (and still can be) a significant source of illness, there would be an overall benefit to avoiding it, thus triggering the nausea and vomiting response.

But what about vegetables? They hardly seem dangerous. The vegetables we consume today have undergone major renovations and do not necessarily contain the levels of plant toxins found in their wild counterparts. Yet even today, vegetables such as cassavas and bamboo shoots contain a plant toxin called cyanogenic glycoside. When consumed raw or undercooked, the chewing action causes an interaction between enzymes and the plant material that releases hydrogen cyanide. Nasty stuff. Toxic.

One reason why nausea and vomiting are particularly effective in helping women avoid foods that could pose a risk to themselves or their developing child, especially if the risk is severe, is that the experience of nausea and vomiting is linked to a powerful one-trial learning experience. We are primed for conditioned food avoidance.

For instance, when I was in my early twenties, I, like many others, waited tables to put myself through school. One stint was at a small Italian restaurant in Delray Beach. Even though I cannot remember the name of the place, two incidents are indelibly imprinted on my brain. The first involved my hand and a hot pizza oven, and the second had to do with a bowl of mussels. I am Italian, so I was no stranger to bivalves. I had eaten my fair share of clams, but I had never eaten a mussel. After eating what seemed like a perfectly fine bowl of mussels, I became violently ill some hours later. I have had food poisoning before—one particular case after eating out in Kathmandu springs to mind—but never after consuming a novel food. That experience became permanently imprinted and I will never try a mussel ever again. No amount of coaxing or reassurance can overcome that primal lesson.

The strength of this response is common in other species as well. Perhaps the most classic and well-known case involves the monarch butterfly. Monarch butterflies are those fabulous orange, black, and white butterflies that float and sail through the air. Adults migrate and their taste for nectar makes them important pollinators for wildflowers. Their numbers are rapidly falling, mainly as a consequence of the monocultures of crops planted in the Midwest, along with the popular pesticide Roundup that kills the food that young larval monarchs need: milkweed.

There are many different types of milkweed plants, but one thing they have in common is they contain toxic steroids called cardenolides. As I already mentioned, many plants contain poisonous compounds. That's because, unlike an antelope, plants can't run away from something trying to eat them. They are kind of stuck, rooted in one spot, if you will. Thus, natural selection has favored an alternate strategy to protect some plants from being eaten: toxic chemicals. Some animals, such as the monarch butterfly, are immune to the negative effects of these substances and can even use them to their advantage. The monarch larvae only eat milkweed, and they are able to sequester the poison, which in turn makes them unpalatable to things that want to eat them. Genius! Incidentally, brightly colored animals tend to be poisonous.

Scrub jays, like many birds, frequently eat juicy caterpillars. But scrub jays that take a bite out of a monarch live to regret it. Scrub jays are large birds belonging to the corvid group (crows and ravens) and are remarkable in their own right. But their elegance is diminished if you catch them barfing after trying to eat a monarch. It's kind of funny in the same way as watching someone walk into a pole is funny. Once you know they are okay, you are free to laugh. It only takes about ten minutes for the jay's digestive system to wholeheartedly reject what it thought would be a tasty meal. And after one experience, lesson learned, most jays simply avoid eating monarchs.

One exposure, that's all it takes. Why? Learning the first time can be the difference between life and death. A food that makes an animal sick might have the potential to kill it or weaken it, making it vulnerable to attack from a predator. This one-trial learning we are wired for is also why an aversion developed during pregnancy could last for years or perhaps even be permanent. That's what happened to my good friend Alma. Alma is one of those women who, if she found herself stranded on a desert island, could construct a makeshift house, build a fire, and cook you a delicious meal within twenty-four hours. But when it came to eating and pregnancy, she wasn't so tough.

What vegetable made her incredibly ill while pregnant with her daughter? Baby corn. A seemingly innocuous vegetable. It doesn't contain any of the potentially harmful toxins listed above. Nevertheless, after a meal out at a Chinese restaurant where she consumed said baby corn, Alma got wickedly sick. She couldn't eat baby corn again for over six years.

If this experience of morning sickness is advantageous, it doesn't seem like it should be exclusive to humans. Ethnopharmacology is a field of study exploring the use of drugs by different animals and human cultures, primarily focused on using plants to self-medicate during times of illness. You might be surprised to know that animals, like us, treat themselves by consuming plants known to have medicinal properties. In other species, a big obstacle to evaluating whether or not individuals are treating themselves is distinguishing between eating something for its nutritional value that also happens to have secondary compounds useful for curing an ailment such as nausea, or deliberately selecting a particular plant that serves no nutritional purpose. For example, gorillas are known to eat a variety of plants that have medicinal properties, including the leaves of a species called *Vitex doniana*, or wild African black plum that is known to reduce nausea.[20] Some women may be familiar with the chasteberry, *Vitex agnus-castus*, used for thousands of years to balance hormones.

Rhesus macaques, next to baboons, are one of the most well studied of the Old World primates. They are also one of the most successful monkeys out there. An early study in the 1970s, one of the few that has examined food intake, rejection, and preferences during pregnancy in another species, found that in a sample of forty pregnant rhesus macaque females, all of them rejected food at a higher-than-normal rate within the first three to four weeks of pregnancy. Unfortunately, there is no way to relate these results to whether the females felt nauseous. There was a distinct pattern coinciding with changes in hormone levels, which we humans also experience and which can cause nausea.[21]

Given the similarity in hormone changes and surges—not just during pregnancy, but also during the menstrual cycle—that influence appetite and other behaviors across a suite of mammal species, it is a reasonable hypothesis that nausea during early pregnancy and food preferences during pregnancy might be more common than we realize in other animals. There is little conclusive evidence, but it's something for my fellow scientists to get cracking on.

Speaking of food preferences, that brings us to food cravings. For some women, the choices seem bizarre, without rhyme or reason.

Barbara, who also happens to be an anthropologist, described the peculiar hankering for potatoes she had while pregnant with both of her children. Any kind of potato. Mashed, fried, baked—all of it. It waned around the end of the third trimester, but up until that point, potatoes please! I would give more examples, but they are as varied as you can imagine.

One craving common to many humans and some animals is for soil, clay, or dirt. This, along with an appetite for eggshells, ice, coal, ash, chalk, and raw starch, is known as geophagy. Instinctively, you might find it repulsive that someone would experience a craving for dirt and follow through with it. Perhaps you think there is something psychologically wrong with the person. That might be true of a subset of people diagnosed with pica—the tendency to consume nonedible items—but while historically geophagy was attributed to a mental condition, research frequently supports the practice. People, and pregnant women in particular, are usually deficient in some nutrients, such as sodium, zinc, magnesium, and calcium. Worldwide, about one quarter of pregnant women consume earth and/or raw starches in the form of cornstarch, raw rice, cassava, etc., and a positive association between these substances and either deficiencies in these nutrients or increased bioavailability has been linked to their consumption, especially iron.[22, 23]

Human females are not alone in their hunger for earth or other substances. Much like the rhesus macaques mentioned above, baboons are another well-studied animal. There are many different species of baboons, including olive, hamadryas, savanna, mountain, guinea, and chacma, to name a few.

The chacma baboon is sometimes also called the Cape baboon and is found principally in southern Africa, including a marvelous place I visited, the Drakensberg mountains. One of the cool things about chacma baboons (and other baboons) is that they show off that they are pregnant—not in their belly, but adjacent to the rough pads they have on their bottom, an area called paracallosal. Males have this area too, but it is only when females are pregnant, roughly two weeks after conception, that the skin color changes from black to crimson. This signal helps not only male baboons know who is pregnant, but also scientists. When researchers took a closer look at geophagy, they found that all individuals consumed soil, but adult pregnant females spent considerably more time practicing geophagy than adult males and juveniles. Among the females, pregnant females engaged in the behavior more than those that were lactating. Also, all the baboons favored sites that had more clay that was higher in sodium.[24]

From the perspective of scientists, be they animal behaviorists like myself, biologists, or primatolgists, geophagy in humans and other animals has been viewed quite differently. Unlike the medical profession, which has traditionally viewed it as a psychological disorder, scientists have observed its ubiquity across species and hypothesized it is a form of self-medication. As a result, they are way ahead of doctors in unraveling this seemingly odd behavior. Aside from delivering nutrients, consumption of clay is known to ameliorate an upset stomach. As we have already seen, pregnancy increases the demand for certain vitamins and minerals (e.g., iron) and causes gastrointestinal distress. Combined, these two factors alone are sufficient to provide a strong biological basis for this behavior without needing to invoke mental disturbance.

Another thing that happens to pregnant women is that many develop a heightened—and even distorted—sense of smell and taste. I can credit my friend Charlotte with my passing biology. Some of you may remember her from my previous book, *Wild Connection*, as the one who had a knack for attracting dogs. Well, she is the sister I never had, and we spent many weekends dancing the night away in South Beach before it was the *it* place to be. When we weren't dancing or waiting tables at Japanese restaurants in South Florida, we were studying hard, and often through the night, to get our degrees and change the world. Unfortunately for Charlotte, during her pregnancies the smell of a café au lait rivaled that of a large pile of dog poop. Given that she lived in France at the time, where she was surrounded by cafés, she found herself constantly muttering, "Okay, who pooped?" during all three of her pregnancies.

All of these mechanisms are in place to protect and foster the healthy development of babies because they are affected in the womb by the environment they experience *through* the parent. The "environment" goes beyond the obvious, such as drugs, alcohol, toxic foods, and other substances that we know are harmful. We are discovering how Mom's (and Dad's) stress or sickness determines gene expression throughout development and even later in life. And we now understand that the dietary choices made by parents influence the long-term health of their children. And if you're a parent, I don't mean what you feed your children. I mean what *you* eat.

WILD LESSONS

* Nausea is an unfortunate side effect of being pregnant, and for some it can be severe. Even though gorillas may self-medicate for nausea, talk to your doctor before taking herbal or other remedies for morning sickness.
* Pay attention to your cravings since they can indicate vitamin and mineral deficiencies. Eat what you crave, except for dirt or chalk! In that case, talk to your doctor.

What You Never Expected to Expect

Beyond food, there are other issues we hear about: peeing constantly, swollen ankles, backaches, discomfort sleeping, increased breast size, and so on. But what about the more unexpected things? Here is a short list of some of the stranger and rarer issues that can crop up during pregnancy:

* Alma had severe leg cramps, her shoe size went up, and she lost the arch in her foot.

- Caroline had obstetric cholestasis, a condition where the flow of bile into the liver is reduced and leakage of bile and bile salts into the blood stream occurs. It goes away as soon as the baby arrives, but aside from causing a potentially dangerous rise in liver enzymes, there is a persistent itch. You itch everywhere, constantly. Enough said.
- Leila had an abnormal gait caused by her hip joints' having become incredibly loose due to the hormone relaxin, which is released to help soften ligaments. As an unexpected benefit, her foot pain vanished.

I could go on, but you get the picture. There are some strange changes that can happen to female animals as well. Things get a tad weird for the panther chameleon. Normally, females are a tawny beige color with a little pink, coral, or teal accent thrown in. Chameleons are known for their ability to change color, but contrary to popular belief, it is not just about camouflage. They alter their color for a number of different reasons, including if the temperature changes, if they feel moody or stressed, or in the case of the male panther chameleon, if they are trying to impress a female.

The mechanism is linked to the complex layered structure of their skin. Panther chameleons, native to Madagascar, have the astounding ability to swiftly change color, especially during social interactions. They accomplish this by activating a network of crystals embedded in tissue layers under the superficial layer of transparent skin, located in another deeper layer of reflective iridiphores that allow them to display and communicate their condition to others.[25] While males are busy changing color to show off their prowess to potential female partners, females put on an altogether different show if they are not free to mate, signaling another male got to the goods first. When pregnant, the panther chameleon female changes color from its usual pale beige to dark brown or black with orange stripes. That is one heck of a signal that she is no longer on the market! Add this to the list—along with the gastric-brooding frog—and breathe a sigh of relief that at least we don't get orange pin-stripes during our pregnancy.

Can I Build My Nest, Please?

Inevitably at some point in the pregnancy process, it is time to get to building that nest. Birds do it, bees do it, even prairie dogs do it . . . and so do we. For some of the women I spoke to, it began early, at the start of the second trimester. Their respective male partners couldn't grasp their sense of urgency because to them the baby's arrival was months and months away. On average though, research reveals that nesting behavior in humans maxes out close to the end of pregnancy and can be bizarrely and narrowly focused.[26]

For instance, Charlotte, the one who couldn't stand the smell of coffee, was obsessed toward the end of her first pregnancy, not with putting together the nursery, but with organizing that one crazy drawer. The second time around, she was compelled to clean behind the refrigerator with a toothbrush. Her husband, Jacques, was mysti-fied when he came home from work to find his very pregnant wife on her hands and knees grumbling about how no one ever bothered to clean behind the monstrous appliance.

Red pandas are rather fastidious nesters as well. An unusual ani-mal, it gave biologists a run for their money when trying to decide just what it was. It shares some similarties to raccoons, bears, and wea-sels. Finally, it was decided that it is so unique it should have its very own genus under the Mustelid superfamily, which includes weasels, skunks, and raccoons. So why in the heck is it called a red panda when it isn't related to pandas at all? I'm not sure, but maybe because it feeds mostly on bamboo, though it is hardly as specialized as the giant panda and is more of an omnivore. And, of course, it looks nothing like a panda.

Anyway, when female red pandas are close to giving birth to their blind and deaf babies, they seek out hollow logs, tree holes, or other crevices to build their nests. But it is about two days before birth that things really kick into high gear for the mom-to-be. She collects twigs, leaves, bark, grass, and moss to construct the nursery. She doesn't just dump it all in a pile either. She carries the material in her mouth and places it in the cavity, then moves it around with her paw, compresses it down with her snout, and heads out for more. She repeats this routine until satisfied and will continue to upkeep the nest as long as her young are in it.[27] Of course, all this nest building, for us and for red pandas, is going on in anticipation of what is about to happen: labor.

Born Up a Tree: When Labor Happens, It Just . . . Happens

In 2001 I traveled to South Africa to visit a friend and colleague living in Johannesburg. At dinner my first night in town, she and her husband raised their glasses and I followed suit, prepared for the "bon appé-tit," or "chin chin," or "cheers." What I was not expecting as a toast was "born up a tree." Huh? And so I was told the story of Sofia from Mozambique. The year prior, in February, there had been a cyclone, Eline, that compounded on torrential rains from just a few months prior that had delivered the annual amount of rain to Mozambique in about two weeks. The additional flooding from the cyclone left many dead and many more trapped, clinging to structures and trees. In the small village of Chockwe, a woman in labor was found in a tree.

As the story goes, the woman had been stuck in a tree for three days when she went into labor. To help, other villagers (also stuck in trees) passed her from tree to tree until she got to the tree where a woman who was a sort of midwife was located, also hanging on, waiting to be rescued. As rescue crews arrived, her daughter was on her way out, unwilling to wait any longer to come into this world. Clinging to a branch a few feet above the flooding, this mother gave birth in a tree.

Holy smokes! Ever since I heard this story, every chance I get, I raise my glass and proclaim, "Born up a tree."

Sometimes labor comes on unexpectedly and sometimes, frustratingly so, it doesn't come at all. That's what happened to my friend Sheila. I have to point out that she relayed this story to me in between breastfeeding sessions and it was a real nail-biter. Here goes. Her pregnancy was blissful, so joyful in fact that she was reluctant and frightened to go into labor, and pretty terrified to be a mom. This is something that no one talks about. It is scary as all get out for some, not only the delivery but also what comes after.

As for many moms-to-be, the artificial due date came and went. When there was no sign of labor and too much amniotic fluid present for the doctor's liking, they made the decision to induce. Remember at the beginning of the chapter how anxiety about pregnancy can keep your period at bay? Well, the same thing can be true of labor. As soon as she was informed that she would be induced—hello, labor! It was mild at first, so hubby went home.

Once she was alone, the contractions began to come on in earnest. When I asked her what that was like for her, she said, "Have you ever had a menstrual cramp?" "Yes," I replied. "Okay," she said, "menstrual cramps are like a firecracker and labor contractions are like a hydrogen bomb." Whoa! And within an hour of being left alone, Sheila's contractions went into full throttle. This seems a common occurrence across the animal kingdom: a large number of animals, including the Zanzibar bushbaby, give birth all alone.

In all my dreams of visiting Africa when I was younger, Zanzibar enchanted me. I imagined a land infused with exotic spices, forests filled with enigmatic creatures, and people dressed in colorful clothing. As I have yet to visit Zanzibar, it still enthralls me. The bushbaby, or galago, is exactly one of those out-of-the-ordinary species you would expect to find on Zanzibar. Bushbabies are a nocturnal prosimian, a type of primitive primate. Moving quickly through the bushes, they use their large ears and eyes to hunt down insects. Normally, females sleep with a male or possibly another female, but in the days before giving birth, females go off on their own.

Much like us, female bushbabies get rounder when they are in the advanced stages of pregnancy, so it is pretty easy to determine when a female is pregnant, at least if you see her up close. One study details the story of three females, one of which was named Mx. P. As is common in bushbabies, female Mx. P normally shared a sleeping site with a male and another female, but on February 21, 1981, Mx. P was no longer sleeping with her companions. Instead, she began to sleep alone—the classic sign of an impending birth. Other than a brief reunion with the male for one night, she continued to sleep alone with her infant for about twenty-four days.[28]

Sometimes, though, females don't go off alone to have their babies. Sometimes a whole group of them will get together and have their babies in the same place. Wildebeest are herd animals featured in so many documentaries for their remarkable migration across the Serengeti in Tanzania, where over two million individuals make their journey to Kenya's lush Masai Mara. What we don't often see is how new wildebeest come into the world. From the migration, one might be inclined to think that wildebeest hang out in these vast numbers, but they don't. Within a given area, they subdivide into smaller groups that are made up of females, or that function as nurseries (females with their young), or that consist solely of bachelors.

There is one other group that forms: pregnant females. The biggest danger to calves are spotted hyenas, so pregnant moms-to-be band together when it is time to give birth to reduce the threat. These "calving grounds" are where births take place. Most births happen in the morning, perhaps also to avoid the dusk and nighttime predator activity. After the sack breaks, the front feet, marked by bright yellow hooves, make their appearance. The best position for birthing females tends to be on their sides, and within an hour, labor is complete— unless it is interrupted. A female would stop the process if (a) a predator showed up or (b) her female companions had moved away in the interim. And so, even with her calf half hanging out, dangling, a mother wildebeest finds being alone so aversive that she will get up and follow the group so as to find company.[29]

For many women, we also either do not want to be alone or cannot be alone. Despite Sheila's rapid progression once she was alone, she wanted her partner there. She needed him there and she ended up requiring an entire surgical team. There she was, in a Swiss hospital, waiting for her husband to rush back when, as she put it, "the evolutionary maladapted part came next." (Yes, this is what happens when scientists have babies. They use this kind of language even in the midst of what proved to be a frightening situation.)

Sheila is *very* slender and she carried her baby high and round. It was the most perfect basketball you could imagine. And it stayed high and round, which means her son never dropped and engaged her pelvis. She had warned the doctors that her family had a knack for making big babies and that her husband came from a line of large heads.

The doctors dismissed Sheila's concerns that her baby would not fit and assured her with the utmost confidence that her pelvis would adjust. That's what women's pelvises do, they asserted. They adjust. Ummm . . . well, you can probably guess that Sheila's pelvis did not adjust. After seven hours of active labor, her son's head proved too large to inspire her pelvis to spread. Add to that her narrow frame and his big body and the end result was labor that would not progress.

Labor is not a walk in the park for many other species, either, and many show considerable distress and pain during the birthing process. Camels, truly remarkable and critically endangered creatures, illustrate this point. There is the one-humped dromedary camel and the two-humped Bactrian camel. (Their humps, if you've always wondered, are filled with fatty tissue.) Incidentally, when a camel gets seriously irritated, it does not spit, as many people believe. That stuff they spray on you is actually a little regurgitated, partially digested food.

So really, they are throwing up on you—just a little. Considering all the other options they have (like kicking you), if you have disturbed a camel enough that it spits *up* on you, you probably deserved it.

These graceful creatures have played an important role in human culture as early as 2500 BC, when the Bactrian camels were domesticated.[30] But they were often depicted alongside horses in rock art drawn by early Saharan herders ten thousand years earlier. The drawings of camels were elaborate and depicted them in motion, often in "flying gallop," probably owing to their speed.[31]

But back to pregnancy. Camels are pregnant for about a year, and calves emerge weighing roughly eighty pounds. About 7 percent of calves die because they come out feetfirst, or breech.[32] For camel moms, labor is intense, and they are known to bellow or hum, which they also do when they are hurt, suggesting that they are in pain. But perhaps hyenas have it the worst. While Sheila had a narrow pelvis to contend with, hyena females must pass their pups through their pseudo-penis, which is really an enlarged clitoris. Okay, men, this is where you collectively gasp, especially if you have ever passed a kidney stone through your penis. I'll never forget sitting on the bathroom floor in a hotel room at the Hard Rock Cafe, holding my friend Oscar as he wept uncontrollably while passing a kidney stone. What emerged finally was a teeny, tiny, barbed crystal. Nevertheless, hyenas must do this, magnified beyond belief, in order for the species to continue. Consequently, there is a very high mortality rate among first-time hyena moms, close to 20 percent.

For Sheila, modern medicine and the emergency cesarean section she received likely saved her and her son's lives. The same was true for Imani, an eighteen-year-old gorilla at the San Diego Zoo, and her daughter Joanne. Imani needed an emergency C-section, and veterinarians and neonatal specialists from UCSD Medical Center worked side by side to save mother and daughter. One of the neonatal specialists remarked that it was because of the similarity between humans and gorillas that things went as smoothly as they did. The biggest difference? Gorilla babies have mad gripping skills with both hands *and* feet. Apparently, this was the greatest challenge when dealing with 4.6-pound Joanne! All went well and Joanne, who is growing up, was reunited with her mom within two weeks of her birth.

This highlights that, despite assertions that only humans have evolved to have difficult and dangerous labor, other animals often run into trouble. For some time it has been proposed that the anatomical changes that facilitated our walking upright resulted in changes to the pelvis that functionally narrowed the birth canal. Combine that with our relatively large head-to-body-size ratio and you have a recipe for what was thought to be a uniquely risky birth process in humans. However, this doesn't really hold up under scrutiny. We only need to look at squirrel monkeys to see that we are not the only ones who need to give birth to babies with excessively large heads.

Squirrel monkeys are fascinating small-bodied primates found in Central and South America. Thanks to excellent vision, they are expert snatchers of insects. They also have, relative to other monkeys, a pretty large brain given their body size. While it remains unclear why their brains are so large, what we do know is that squirrel monkeys have one of the longest pregnancies of all New World monkeys (those found in South and Central America) and most Old World primates (those found in Africa and Asia). It is possible that this long gestation period is needed for their brain to grow and to bring baby squirrel monkeys up to speed so that they are not completely helpless, unlike us, when they are born. It also allows them to mature very rapidly, because their brains are already mostly fully developed at birth.[33]

What this means for female squirrel monkeys is that they can have a hard time giving birth, and so the infant's head reorients to pass through the pelvis face-first. Basically, the infant's head bends to accommodate the length, but it cannot do anything about the *width*, so the cranium is shaped a bit longer than normal. But it is still 121 percent wider than the birth canal. It is the narrow distance between the eyes that allows the infant to fit exactly—most of the time. (Many moms and babies still die during the process.)

It is thought that we humans have a trade-off: the gestation period and energetic demands required to accommodate maturing our brains so that we are more precocious at birth would result in a huge metabolic cost to the mother and a head way too big to ever fit through the human pelvis, even if our pelvis hadn't undergone the structural changes to allow us to walk upright. But wait. What body part gives us the most trouble? The shoulders, which are a birth-canal conundrum and are responsible for more complications than the head because they are quite inflexible.[34]

The struggles that squirrel monkey moms face have helped overturn the notion that only humans have a hard time in labor and experience death, pain, and anxiety. The high mortality of squirrel monkeys and spotted hyenas, the sweating of mares, the restlessness and humming of camels are more than sufficient evidence to supplant this idea. It also contradicts the assertion that only human infants must undergo "fetal rotation" (turning) to pass through the birth canal. Other animals also turn, just sometimes in the other direction! In other primates, the back of the baby's head fits better at the back of the pelvis, while in humans it fits better in the front.[35]

As for the idea that only humans experience worry and nervousness, this ill-conceived suggestion is based on the premise that these feelings are cognitive in nature, ergo only human. But this woefully ignores the complex cognition of other species, which we are coming to understand more and more. Granted, our penchant for *obsessively* agonizing over every little detail of what might happen many months (or years, for some people) down the line could be uniquely human, but the scientific literature is rich with evidence that other animals experience fear and anxiety in anticipation of events that cause pain.

One of the reasons it has been suggested that only human mothers are nervous and frightened is that in the majority of cultures a woman receives assistance during childbirth. We saw earlier that wildebeest give birth together, to the extent that a mother will get up during labor and move to be with others. And while wildebeest do not get actual birthing assistance from their fellow wildebeest friends, humans are far from the only species that does. It is fairly common in marmoset, tamarin, and titi monkeys, with dads helping moms throughout the birthing process.

The same is true for the Djungarian hamsters, sometimes called dwarf hamsters, which are tiny bundles of cuteness that fit with room to spare in the palm of your hand. In this species, the pair bond is strong and males provide paternal care. But they don't start after the babies are born; they get right in there and get their paws dirty. Literally. While moms are in labor, male Djungarian hamsters help by licking amniotic fluid, using their incisors and paws to reach in and help deliver pups, clear the nostrils and airways of pups, completely clean the pups off, and finally carry them to the nest. All during this time, the female is left to deliver the placenta and clean herself up.[36] Go Dad! This assistance, just as with other species where both parents are needed, helps the pups survive.

In the case of Rodrigues fruit bats, it isn't the dad but a sort of midwife fellow female fruit bat that helps a mother out. This species lives only on the island of Rodrigues, in the Indian Ocean, and their population is precarious. Human hunting, habitat loss, and cyclones represent major threats to their continued existence. Like many bats they are very gregarious and roost together in large groups. In most species of fruit bat, it has been observed that baby fruit bats are born headfirst, while most echolocating bats are typically born feetfirst.

Researchers observed the birth of a bat in a captive colony containing unrelated females that were on loan and had no previous interactions. In other words, they were strangers. It took a full three hours from start to finish, and what they witnessed was remarkable. The helper female not only helped by licking the anogenital region of the mother bat that was in labor, but she guided, using her bat hands (since bat wings are really modified hands), the mother into the proper birthing posture! Normally they roost head-down, but they need to be feet-down to give birth. In addition to providing direct assistance, the bat acting like a midwife gave the mom a demonstration by getting in full view and flipping over.

It was only after the mother-to-be received this assistance that she reoriented into the correct position. Once in the right posture, the helper bat would go back into a head-down position and wrap her hands around the mother. As a baby bat's foot emerged, the mother assumed the cradle position, and the helper continued assisting until two feet and a wing came out and the baby bat was able to grasp onto its mother's foot with its thumb.[37]

Not all bats, hamsters, monkeys, or humans need help, but sometimes it is essential. Without the cesarean delivery, Sheila and her son may not have survived. This has been the case for females of many species around the world that bring new life to this glorious planet. Many humans, and other species in captive environments, have the benefit of medical intervention if needed. Among humans, medical advances have certainly reduced mother and infant death rates in many places. However, there has been a disturbing upward trend in mortality during childbirth, with the United States ranking very high among developed countries, averaging almost eighteen deaths per 100,000 births.[38] There are a myriad of culprits, but some doctors believe the advancing age of women becoming pregnant, increased rates of non-emergency cesarean sections, obesity, and growing health disparities are largely to blame. For us and other animals, these are the risks we must take. But, not to minimize the real dangers that some women face giving birth, thank goodness we are not spotted hyenas!

WILD LESSONS

* When it comes to labor and potential complications, pay attention to your instincts.
* Even though labor and delivery is a "natural" process, that doesn't mean it is without risk for us and other animals. Worry, anxiety, and fear are common feelings experienced by humans and other animal moms.
* Whether it is your male partner showing up like Djungarian hamsters or another female assisting like a Rodrigues fruit bat, many women welcome and need help during labor. Ensure that you have the proper support system in place to have a successful delivery, even if you find yourself stuck up a tree!

CHAPTER 3

Your Bundle of Joy

I hardly think that anyone would argue with the assertion that bonding between newborns and their parent(s), biological or otherwise, is of paramount importance in the development of healthy children. Because of this, recognition between parents and their newborn offspring and vice versa is a salient feature of forming this connection. Whether you are a spiny mouse, a goat, a marmoset, a seal, or a human, making and taking care of babies is a substantial investment, one that involves interaction with and care for offspring that are usually your own. Thus, we, and other species, have multiple mechanisms by which we can identify, often rapidly, which baby is ours and bond accordingly. If any one of these systems fails, there may be a backup, a fail-safe if you will, intended to preserve this system of mutual recognition and affection.

Sometimes, as in my case, all of the backups crash. Though the details have never been made clear, what is unmistakable is that my birth was not without incident. My mother had to undergo emergency surgery immediately after my birth and for approximately three minutes she was clinically dead. She did not hold me or even see me for some time, nor did my father. I was sent home with my grandmother while my mother recuperated in the hospital. Later, once my mother came home from the hospital, it was still my grandmother who exclusively cared for me for several more months. Undoubtedly, this was the reason why my bond with my grandmother was so strong. When our biological parents are unavailable to care for us, is it not our saving grace that we, like so many other species, are born cute, thus evoking the caretaker response in others?

By now you are likely aware of my passion for gorillas, so I love this anecdote about a gorilla who reacted to the helplessness of a young member of a species not its own. In 1986, a five-year-old boy by the name of Levan Merritt would forever change our image of gorillas. The movies would have you believe in a violent King Kongish silverback, though gorillas are arguably the most peaceful of all the great apes. Jambo, a silverback male living at the Jersey Zoo (now called the Durrell Wildlife Park, founded by author and naturalist Gerald Durrell) made famous this gentleness in a gorilla's nature.

Levan had fallen into the gorilla exhibit, smack into Jambo's territory, with Jambo's females and his own children. Levan was knocked out from the fall. In the video, the visitors gasp and some begin to cry as Jambo slowly makes his way over to the unconscious Levan. Once near Levan, who is lying facedown, Jambo places himself between the boy and the other gorillas in his family, protecting him. And then he sits next to Levan and rubs his back, as if to comfort the boy.

Jambo recognized that Levan was a young human boy and protected and cared for him accordingly. The ability to recognize youth and helplessness that requires care is quite possibly universal. The hallmarks of "youngness" are all over us and other species. As you will see, the fact that kids are so darned cute is not an accident.

Furthermore, for us, and for many other species, care is usually provided by one's parents, and the bond between parents and their offspring is forged early, often in the first few hours of life. In my case, I was always more closely bonded with my grandmother, Oma. Was my deep connection to my grandmother due to hers being the voice I came to know, the smell I was familiar with, and the touch that soothed me in those first critical moments of life?

Aren't I Cute?

How many times do we hear the phrase, "Oh he/she is so cute!" in response to seeing a baby anything? Okay, maybe not baby snakes, though personally I even find baby snakes rather lovable. In Spanish it's *qué lindo*, while in Japanese it's *kawaii*. In whatever language one speaks, the words alone evoke a smile. Universally we respond to cuteness.

The cuteness *can* take a little while to kick in. Though I have seen some rather adorable newborn photos, I think we can all agree that, for the most part, human babies do not look cute the moment they are born. They look slimy, alien-like, with large, somewhat squished heads, fluids oozing out of their nose and mouth and covered in . . . well, let's just stop there.

But once human infants, and other species, get past that squashed newborn look, they are simply, deliciously irresistible. When I see a baby animal, I almost cannot help myself from holding it, cuddling it, smelling it, feeding it, nurturing it, keeping it, and finally, making it mine. My rational brain resists this last urge most of the time—I say most of the time since I do have three animals in my household, the result of falling madly in love with two very cute feral kittens born on my doorstep and adopting them and their mom.

Even baby alligators are disarmingly cute. For a few months I volunteered at the Loxahatchee National Wildlife Refuge in South Florida, where much of our time was spent surveying various species. By day we counted egret nests or dealt with exotic species abandoned in the Everglades by their foolish human owners, and by night we caught and measured baby alligators. They have heads much larger than their bodies, enormous eyes, and they make the sweetest sounds. Might this explain the lapse in judgment in people who think it's a good idea to have an alligator as a pet? Like people and other animals, baby alligators grow up, progressively becoming less cute.

Given that almost all babies, regardless of the species, evoke a similar sentiment, it made me wonder what all this "cuteness" is about. Does it serve a purpose or is it an accident that infants of all kinds, including humans, are all eyes, head, and round, soft, clumsy bodies? Konrad Lorenz, one of the founders of animal behavior science, with a thing for greylag geese, was first in tackling this question back in the 1940s and '50s. Lorenz made the argument that "cute = baby" and that the exaggerated features of babies serve as a signal, triggering adult animals, including humans, to automatically behave in a nurturing and physically gentle manner toward the young, while simultaneously suppressing aggression.[1]

For example, when an infant cries, many of us feel concerned, even uttering, "Awww, what's the matter?"—unless it's at three o'clock in the morning, on a plane, in a restaurant, or during a movie. With those caveats, have you noticed that once a human baby passes this early-infant stage and is stumbling about in characteristic toddler fashion, we are far less forgiving of such offensive wailing? This initial "awww" is there to protect offspring of all species in their most vulnerable phase. Although, truth be told, I am more likely to give an adult panda more leeway for fussing than any human child over the age of two. Perhaps that is because even adult pandas look cute with their large heads and those eye-widening circles.

If infants looked like something straight out of a sci-fi movie, they might be at considerable risk from adults' aggression with all the crying, begging, and squawking they do. Protecting infants from aggression, then, is facilitated by all these physical hallmarks of infancy: large heads, big eyes, and wobbly, soft bodies. Does it really work? Are we actually wired to be more caring toward things that look like infants?

We just might be. In one study, the researchers were interested in determining not simply how emotionally responsive people were, but also how physically careful they became after viewing images of young and adult animals such as dogs and domestic and wild felines (e.g., lions and tigers). Remarkably, the study found that just looking at pictures of baby animals is enough to make people, both men and women, more cautious and use a gentler grip while playing Operation.[2] Who could forget that childhood game that demands more coordination than is natural for a seven-or-eight-year-old? Apparently, all that is required to improve your "game" is to look at pictures of babies. Go figure. And it's not just empathy and caution that are enhanced. A separate experiment revealed that concentration and focused attention are substantially increased just by looking at photos of baby animals. In the above study, after looking at images of baby mammals, human test subjects were able to find a specific number in a random sequence of numbers more accurately *and* faster than people who looked at photos of adults or food. So, basically, we are more careful with things that are cute.[3]

What if a baby isn't cute? It happens. There—I said it. Some babies get cute as they get older and some start off cute and just get older, and some, well, you know. The important question is whether or not it will influence how a caregiver treats an infant. Or even how you treat your own child. There is some evidence to suggest that when an infant is rated as unattractive, he or she is judged to be older (and therefore *more* capable) but simultaneously less developmentally skilled.[4] This means that when we look at a less attractive baby we think that baby *should* be able to do more but is not proficient enough to do so. A contradiction, to say the least. And this goes for parents as well as unrelated caregivers. Aside from a slew of judgments made about the age and competence of attractive versus unattractive infants, how attractive an infant is may just predict how much attention and interaction that infant gets from his or her mother and other caregivers.[5]

As we already witnessed with the bluebird dad's preferential treatment of his more brightly colored son in Chapter 1, this preference for cuteness is not unique to humans; there is a difference in parental investment based on perceived offspring quality in other species as well. Basically, parents are predicted to pay more attention to offspring they determine have a higher likelihood of future success. And like it or not, in humans attractiveness can be one indicator of success. Studies repeatedly reveal that more attractive people get hired more often, get paid more, and date more than unattractive people. As I discussed in *Wild Connection*, within one tenth of a second we decide whether a potential mate is attractive, competent, kind, and friendly. Why does this also happen with our babies? It might just be because parents divert a tremendous amount of energy toward their kids, and, biologically speaking, you want all that effort to pay off.

The good news is that this bias—namely, how much attention moms or dads give to less attractive infants—theoretically declines with increased bonding time. But, as we will see in Chapter 6, playing favorites doesn't go away, and parents can and *do* allocate time, energy, and resources unequally among their offspring.

But first, back to cuteness and its effect on adult behavior. Besides the big heads and oversized eyes, humans infants and young of other species have a slew of features that trigger the same gentleness and tolerance, not just in the parent, but also in other individuals, both male and female. Have you ever looked closely at an infant or young chimpanzee's bottom? No? Something for the bucket list then. If you have, then you surely noticed the little tufts of white hair placed squarely on their backsides. It's a signal to everyone in the community to be more tolerant than they otherwise might be. This signal protects the infant from aggression, even in the face of a serious social faux pas.

WILD LESSONS

* Humans are more physically cautious with infants because of the "cuteness" factor.
* Spending time with your infant overcomes any unconscious bias based solely on physical appearance.
* Be aware that others may alter how they treat your infant depending on his or her appearance.
* Discuss any potential bias in care with you childcare providers.

This innate mechanism that triggers parents to respond to the exaggerated stimuli characteristic of babies may also lead some species to take advantage of the parenting instincts of others. Like cuckoos. Some species of this bird are called brood parasites because they don't bother taking care of their own kids. Instead they rely on triggering the strong parental instincts of other birds that then raise the cuckoo chick—more often than not to the detriment of their own offspring.

The cuckoos that engage in this practice may lay their eggs while the unwilling adoptive parents are away from the nest. Others, like the great spotted cuckoo, are not above shoving a sitting magpie right off the nest and dumping an egg right then and there.[6] These cuckoo "parents" tag-team, allowing the magpie parents to launch an all-out attack. When the right moment arises, the cuckoo female quickly lays an egg and off they go. Then, like many other species tricked into raising a cuckoo chick, the magpie goes right back to incubating her eggs, including the cuckoo's. Are the birds ever able to detect a stranger's egg in their midst? Some species parasitized by cuckoos have fought back by evolving complex and distinctive egg patterns as a means of distinguishing between their eggs and cuckoo eggs.[7] Not so for the magpies. They are tricked into preferentially feeding a cuckoo chick by the larger size of the chick as well as the bright color, which acts like a beacon as it gapes, begging for food.[8]

Are You Mine?

There is a difference between looking cute and looking like your parent. Whenever a new human baby enters into this world, the first thing that everyone seems to say after "adorable," "beautiful," "cute," or some other synonym is, "Oh, she/he has your eyes!"—or nose, or chin, or some other body part. Sometimes it is true and the baby is a dead ringer for mom or dad. Other times, even the parents remark, "Hmmm . . . you really think so? I can't tell." Where does this obsession with confirming that a baby looks like the parent, especially the father, come from?

Whether you have your baby alone or in a crowd, it pays to be able to distinguish your kid from someone else's. As behavioral ecologists, my colleagues and I often like to think in terms of costs and benefits, so let's review the costs of failing to recognize your own offspring.

Immediately two things come to mind. First, if you don't know your offspring is yours, you are highly unlikely to take care of him or her. Without food, shelter, and protection, any newborn that requires at least some parental care will likely perish. As if that isn't bad enough, not only might your offspring die, but you also might end up diverting your precious resources to someone else's offspring, even if inadvertently, as we saw with the magpies.

Given this, it seems the fact that magpies sometimes cannot recognize that an individual in their nest is not their own is an unfortunate outlier. For the majority of nesting birds, the default setting is to think, *If you are in the nest, you must be mine.* And if you are not beleaguered by cuckoos forcing you to foster-parent, or if a mate has not committed infidelity (always the mom in the case of these birds), this rule of thumb would work 100 percent of the time.

Given that this detection rule does not work for an abundance of other species—and, of course, recognizing egg patterns or nest occupancy is not a viable option for humans—it should come as no surprise that there is a suite of alternative mechanisms by which humans discern just which one of the crying baldies in the hospital nursery is theirs, and that others do essentially the same for their own young. Keep in mind that from the infant's perspective, he or she is not invested in his or her own parents, merely in getting the required care. So the onus is on the parents to figure out who is who. The obvious method is to base this on what your child looks like and, more importantly, how similar she or he looks to you or other family members. This is called *phenotype matching*, where phenotype represents the external appearance in any given trait (for example, nose, eyes, chin, hair color).

When it comes to humans, and all placental mammals really, it is pretty obvious who the mother is. Who is the father, though? That can be up in the air sometimes, so it should come as no surprise that men use perceived facial resemblance as one cue to drive their paternal affection and investment. Instinctively, many a mom will say to her husband, "Oh, he definitely has your eyes!" reassuring her partner that this indeed is his child. The reality is that men are able to discern their offspring at a rate higher than chance alone.[9]

Aside from paternity assurance, facial resemblance enhances trust between any two individuals.[10] When we perceive that someone looks like us, that person is instantly familiar and this "sameness" in appearance changes in a positive direction how we interact. The fact that we are primed to trust individuals that look like ourselves has many implications beyond parent-offspring recognition. However, the need to recognize one's own offspring is so fundamental to reproductive success that the origin of this phenomenon is easily understood. And it doesn't just happen in humans.

Like many colonial nesting seabirds, ring-billed gulls nest among thousands of others. Everyone is scrapping for their little piece of terra firma, and there are frequent altercations with neighbors. This means that parents and chicks can get separated, necessitating finding and recognizing each other. One way the ring-billed parents do this is based on what their chicks look like.

To test this, in one experiment, researchers modified the appearance of the plumage and facial characteristics of some hatchlings. Given the distress that chicks (and humans) experience when their parents reject them, the results of this experiment were pretty upsetting. When a ring-billed gull parent encounters a chick they *believe* is not their own, they peck at it and treat it roughly, even evicting it from the nest. And when the parents in the experiment encountered their own chicks with the facial and plumage features altered by the experiment, they frequently attacked them and rejected them, despite their chicks approaching them in the appropriate *I'm yours* posture.[11] Basically, Mom or Dad didn't recognize them anymore and treated them accordingly. Fortunately, the experiment was short-lived. The changes in appearance were temporary and all went back to business as usual in a few days.

We humans are pretty adept at discerning our infants solely on the basis of a photograph shortly after they are born and with a few hours of contact. When new moms were tested, twenty-two out of twenty-four correctly picked out their infant from a lineup of newborns that all looked similar.[12] It remains unclear exactly what cues the moms used (for example, the individual characteristics of their baby or familial resemblance) to identify their infant. Even though fathers weren't tested in this study, I suspect they would have scored similarly.

Let's Imprint

But let's face it, not every baby, whether a human or a zebra, necessarily looks like its parent or even looks unique. So if you can't tell who your baby is by looking at him or her, how are you supposed to know? Indeed this is a quandary faced by parents of multiples who are identical. Twins, triplets, or even more multiples of identical babies present a unique challenge for new parents. You might easily name them in the hospital and keep getting them mixed up until they are older and start showing some more obvious differences. To avoid this type of confusion, some parents resort to color-coding their multiples' clothing or even painting their toenails!

Although many animals have more than one offspring at a time, they may not necessarily need to recognize their offspring individually. They must, however, have a way to tell their babies apart from someone else's baby. Since animals can't color-code their batch of babies, the next best (and very common) strategy is to do what geese, ducks, swans, cranes, chickens, and many other species do: Have their babies imprint on them. For any one of these species, this means that the first suitable parent or set of parents that a newborn sees becomes the parent. And we all know what baby chickens do—they follow the individual they've determined to be the parent, relentlessly.

Greylag geese get their name because they migrate the latest in the season, thus "lagging" behind all the other species of geese in migration. But when it comes to long-term memory and understanding social dynamics, these geese are anything but slackers. When you look at geese, your first thought may not be of a complex social system, but because greylag geese form long-term relationships with a partner and have deep-rooted family associations, they have a lot to keep track of. And flocks of geese are not willy-nilly aggregations of random individuals. They often consist of multiple family units, and there is a pecking order among members of the flock, with family units looking out for each other. For all their social sophistication, when it comes to imprinting, they are less than discriminating.

As Dr. Lorenz discovered, a newly hatched greylag goose will imprint on *any* moving object. It doesn't even have to be alive. It is as if the inexperienced brain of a baby goose says, "Something is moving—that must be Mom and Dad!" That hardly seems beneficial. Imagine if, within thirteen to sixteen hours of being born, you were not with your infant, but your family cat was, and your newborn imprinted on sweet Milo.

So what gives? Why do some species, such as chickens and greylag geese, imprint visually and instantly on their parent(s)? One thing that all these species have in common is that they are precocial. Simply put, this means that shortly after birth, or hatching, they can get up and follow their mom or dad around. Their mom and/or dad don't feed them so much as guide them around to find food and offer protection from other adults and predators.

Many other species, including humans, are altricial, or essentially helpless. In that case, a system like visual imprinting, though it may certainly occur, wouldn't work as the *primary* mechanism. An infant human or tree sparrow can't follow its mom and dad, so there have to be other ways for offspring to recognize parents and, more importantly, for parents to recognize their offspring. One alternative might be vocal recognition.

You Sound Familiar

Imagine for a moment that you are a Mexican free-tailed bat parent and you live in a crèche with hundreds of thousands, possibly millions, of other Mexican free-tailed bats, crammed together, and you all look—well—sort of alike! Incidentally, if you've ever been to Austin, Texas, you may have visited the Congress Avenue Bridge, which houses one of the largest colonies of Mexican free-tailed bats, estimated at well over one million. And before you dismiss bats as weird, gross, frightening, or dangerous, these particular bats eat insects—thousands upon thousands of pounds of insects. If you're a farmer, it pays to have bats around. They are better than any pesticide, and they're known for being economically valuable to the Texas cotton industry.[13]

When it comes to taking care of their pups, Mexican free-tailed bat moms overcome an insurmountable feat: recognizing their offspring in a sea of millions of other pups—as many as five thousand crammed within every ten square feet, which is just a bit smaller than the average American guest bedroom. How is a mother bat to solve this problem and ensure she can recognize and feed *her* offspring? Her first task is to remember where she left her pup; she then returns to within one meter of that location. Then she must recognize the cries of her baby amid the cacophony of other pups.[14]

Fur seal pups are adorable. They are also noisy. Well, seals in general are pretty talkative. Every year, like other species of seal, the subantarctic fur seals haul out on breeding rookeries, where all the females give birth at roughly the same time. Males are around looking for their next mating opportunity and rarely leave the beach or rocky shore for the duration of the breeding season. Females, once they give birth, must continue to acquire enough food to fatten up not only their darling pups but also themselves for the next pregnancy. What this means is that after about a week, the mom leaves her pup alone and, as in the case of the Mexican free-tailed bats, the young pup mingles among many other seal pups and may not be exactly where its mom left it. It is particularly crucial that moms are able to recognize and reunite with their babies, because these seal moms will reject and sometimes attack strange pups that try to nurse from them. Of course, all of the seal pups are pretty hungry after being left alone for several weeks, so who can blame them for trying to get some milk from any available lactating female.

So how do moms find their pups? Although fur seals have excellent vision, they are nearsighted when on land. Therefore, if you're a fur seal mom, looking for your pup isn't the best strategy unless you're practically on top of it. Just like the Mexican free-tailed bat, mom subantarctic fur seals develop the ability to recognize the sound of their pup's calls—within a few hours after giving birth! Since they don't leave their baby's side for almost a week, their pup also learns the sound of its mom's call within a few days. The end result? Typically, mother and baby are reunited within eleven minutes of separation.[15]

In the early days, newborn subantarctic fur seal pups' cries are high-pitched and quivering, but as they grow and develop, their calls change and the moms have to keep up, continually learning to recognize their calls as they mature. We will talk more in a moment about the role of smell in identifying offspring, but fur seals do have a great sense of smell on land, which is useless in the water, since when they swim they keep their eyes open and their nostrils shut. However, it seems that the moms only smell their babies as a final check, not as the primary way of recognizing them.

What about us? Can any of these strategies help us tell our newborn baby apart from other infants? Sure, there might be visual recognition and phenotype matching, as mentioned above, but if this were a perfect system, there would never be any babies switched at birth unbeknownst to the parents. Back in the 1960s, scientists got reports from new mothers who, while in hospital rooms shared with other mothers and infants, claimed that they were awakened only by the sounds of *their* infants. Skeptical of these reports, the scientists decided to put them to the test.

They evaluated maternity wards where some mothers had rooms to themselves and compared them with mothers who were in rooms with three other new mothers. They recorded the cries of newborns within fourteen hours to six days of birth. Each of the mothers alone in maternity rooms was then asked to listen to five crying infants, one after the other, presented in random order. Only one was the mother's own infant. Each mother was then asked to identify her baby. For the moms in maternity rooms that had multiple mothers and newborns, the scientists recorded how many times each mother was awakened by crying that came from her own infant.

Although fur seal moms recognize the calls of their newborn pups within hours, only about half of the human moms could correctly identify their infants' cries from other infants prior to forty-eight hours. After forty-eight hours, however, accuracy soared to almost 100 percent. Similar findings were reported for the mothers who were awakened by cries, where forty-eight hours after delivery, mothers reported being awakened almost exclusively by their own infants' cries as opposed to the cries of other infants in the room.[16]

As I've already mentioned, from an evolutionary standpoint, it seems pretty self-evident that there are huge benefits to recognizing your own offspring and enormous costs if you fail to do so. If you're a new mother, constantly waking up to the cries of someone else's infant would deprive you of much-needed rest and, worse, cue you to provide precious milk, or "liquid gold," as it's sometimes called, to another individual's offspring. For the most part, mothers today are not at risk of feeding and caring for the wrong infant, but historically this may not have always been the case, thus necessitating a fail-safe mechanism for identifying our own offspring. In some sense then, we, like fur seals and Mexican free-tailed bats, vocally imprint on our babies, and our babies have a unique vocal signature right out of the gate. We can learn this signature whether we are biological or adoptive parents.

So far we have only been talking about how mothers identify their newborns based on sounds, leaving out dads. Biparental care is seen in 5 to 10 percent of mammals, but is much more commonplace in birds. Razorbills are colonial breeding seabirds, related to puffins, that spend their life at sea except for when it's time to have a family. Razorbill relationships usually last for life, and once they pair up, they typically have one chick per year. Their modest nests are a collection of pebbles and other materials placed in crevices with easy access to the water. In the early stages, just after the chick is born, both the mom and the dad take turns heading out to sea to bring back food for their offspring. The task is shared fairly equally, and the chick is never alone. Consequently, if you're a razorbill, there isn't a lot of pressure to try to figure out which chick is yours. Presumably, if you can find your nest and recognize your life partner, then it follows that the chick in the nest is yours.

But everything changes after about three weeks. At this point, baby and dad razorbill set off to sea, where dad is now the sole caretaker. There is also no longer a nest to call home, and father and chick could easily become separated. Because of this twist in parental care events, it turns out razorbill fathers are much better at recognizing the calls of their chicks, even while they are still in the nest. Compared to the moms, the dads respond more frequently and are more adept at discriminating the calls made by their chicks from the calls made by others.[17]

How can human dads become as intimately familiar with their child's cries? It's pretty simple. Be involved. When human dads are just as involved as moms, they too are as sharp as razorbills at distinguishing which cries are coming from their own infants as opposed to other infants. It is not clear whether they can identify the cries coming from their infants as quickly as the moms, but among parents with infants less than one year old, there is no difference between mothers and fathers in the ability to correctly identify which sound is coming from their child. The deciding factor is how much time parents spent with their infant. For both mothers and fathers, spending less than four hours a day reduces accuracy by almost 25 percent.[18]

And it's not just the parents who learn to identify the sounds of their babies within hours or, in the case of humans, two days; infants also rapidly learn to recognize the voices of their parents. Being born as a baby bottlenose dolphin comes with some challenges. First, they need to swim immediately and breathe. Also, because they are born into a large social network of other dolphins, the chance that a newborn could get separated from his or her mom is pretty high. For this reason, mother bottlenose dolphins are rather possessive about their newborn calves and try to keep some distance between themselves and the rest of the group, at least for a few weeks.

As soon as they are born, dolphins are able to produce the characteristic dolphin "whistle" we've all come to know and love. Just as all of us have our own voice, recognizable to friends and family, each dolphin develops a signature whistle—almost like each dolphin has a name. This is how researchers have discovered that dolphins gossip. When two dolphins include in their chatter the whistle of an individual that isn't present, it would appear they are "talking" about another dolphin that isn't around.

The trouble with voice imprinting for mom dolphins is that it takes a few weeks for newborn calves to find their own voice, so to speak. This means that, unlike with other species, where the mom can learn the sound of her baby quickly, it would take a few weeks for a dolphin mom to be able to do this, since her calf doesn't have a unique sound until then. So, to ensure it doesn't get lost, a baby dolphin learns the sound of its mother's whistle! To facilitate this, the mother increases her whistling rate tenfold as soon as her calf is born.[19] Perhaps this is why moms-to-be instinctively start talking to their babies before they are even born.

Since the 1980s, we've known that human infants prefer the sound of their mother's voice. As helpless as we and many other species are when born, it is utterly remarkable what we can accomplish within hours of coming into the world. In one study, ten newborns were subjected to a voice recording of Dr. Seuss's *And to Think That I Saw It on Mulberry Street*. They could choose a recording of their mother's voice over that of another female by suckling harder on an artificial nipple. The more they suckled, the more the recording played; thus they had the ability to "control" hearing their mother's voice over someone else's. Within a twenty-minute training session, the less-than-day-old infants learned to produce the sound of their mother's voice more frequently.[20] And all of these infants had less than twelve hours of out-of-the-womb contact with their mothers prior to testing! This is pretty remarkable considering it took new moms almost forty-eight hours to learn the sound of their infants' cries. There is also evidence suggesting that, within thirty hours of being born, human infants can already discriminate between their mother's language and a foreign language.[21]

WILD LESSONS

* With a few hours of physical contact immediately after birth, mothers (and probably fathers) can recognize their infants just by looking at a photograph.
* Human infants have a vocal signature that new moms can identify within forty-eight hours.

continues . . .

- For a mother and father to continue to recognize the sound of their infant, they must spend more than four hours per day with their infant. If not, accuracy drops by approximately 25 percent.
- Like dolphins, human infants imprint on the sounds of their parents. Pregnant moms can facilitate vocal recognition by talking to their infants while still in utero.
- It is not advisable to play other music or sounds via headphones placed on the stomach. Not only are babies learning mom's voice about ten weeks prior to birth, it is also too loud!

What Is That Smell?

What is it about the smell of a baby? Except for their poo, babies smell fantastic. They smell so delicious that someone even tried to bottle the essence of the smell and sell it. The product was called Baby's Head Smell—Lavatory Mist. It has been discontinued (perhaps it didn't live up to its name?) but clearly demonstrates the popularity of *eau de bébé*. Science backs this up, too. The reward center of the brains of both mothers and non-mothers exposed to the scent of an unfamiliar two-day-old infant light up like firecrackers.[22] And just as human fathers can be adept at recognizing the sound of their infants' cries, they demonstrate a strong response to the smell of their own infants. Both mothers and fathers can pick out the odors of their own infants and these olfactory cues may, in part, trigger parental behavior.[23, 24]

And it's not just human babies. I truly savored the scent of the infant chimpanzees I helped care for many years ago. And honestly, puppies smell divine. Well, except for their milk breath. Have you ever watched people hold a baby? Almost instinctively they nuzzle their head into the neck and inhale. What is this all about? Why are we compelled to smell our babies (and other babies)?

Right out of the gate, we, like other mammals, have a well-developed olfactory system primed to smell. Actually, it begins in the womb.

For newborns—be they rat, rabbit, or human—smelling their own amniotic fluid provides a calming effect.[25] Even though human newborns can theoretically see, their eyesight isn't developed enough to put everything in focus. Essentially, when we are born we have to rely more on smell and sound than on vision.

One strange tradition of humans in modern times is that, unlike other species, human babies are frequently immediately wiped clean by other people soon after they are born, which removes all of this wonderful, calming, familiar smell that helps them adjust to this crazy, loud, bright, strange new environment. Much as we all have a signature voice, we all have a signature smell. That means that the mother has one smell and the baby has another, and we are biologically wired to recognize each other based on these smells. And as I have already mentioned, the recognition between a parent and its newborn is paramount. When we wash newborns off immediately, we are removing a fundamental mechanism that not only calms the infant, but also may disrupt the parent-offspring recognition process we desperately require.

I should point out that in Western societies this practice is changing and is highly variable, depending on whether it is a home-assisted birth, a birth overseen by a midwife, or a birth in a hospital setting. Today, even some hospitals are changing how they handle infants as additional research is gathered. There is also a tremendous range of individual preference in how mothers- and fathers-to-be view this practice. For the few parents-to-be out there who are concerned about holding their baby only after it has been thoroughly "de-gooed," I humbly submit that this apprehension seems misplaced, considering that there will be plenty of unpleasant substances you'll have to deal with soon enough—ones that don't play a critical role in a child's healthy development.

Which brings us back to baby poo. Recently, I was at the one-year birthday bonanza of a friend's daughter. There were at least five other babies less than one year old. The scene was just what you would expect it to be: the parents chattering about their babies, the babies oblivious to the fact that they were indeed at a party, and the childless among the group left to entertain themselves. This put me squarely in the sights of a seven-year-old boy. He and I forged a solid bond when we were faced with the unmistakable and unique odor of . . . baby shit.

We further communed in our mutual disgust over what the parents did upon detecting this distasteful odor. All chatter ceased and, without exception, every parent (mother and father) lifted their baby and buried, I say *buried*, their nose into their child's diaper and . . . inhaled. I recognize that this was performed in an effort to detect if it was their baby that needed changing, but still. . . .

Then my friend Lisa reminded me why I shouldn't have been so horrified. She was quick to point out two things. First, you have to deal with feces every day as a parent, so there is the habituation factor. Essentially, you must quickly overcome your natural disgust because, well, for the most part, your baby is going to defecate every day. Second, if a baby is wearing a onesie—and most babies these days seem to—there is no other way to quickly gauge if he or she needs to be changed. Touché.

Despite her reasoning, I still think that stinky diapers are more unpleasant than amniotic fluid covering your newborn. That is not to say that the placenta isn't removed or that the amniotic fluid isn't cleaned off in other species. It is. But, there is a crucial difference. The parents of other species typically lick off the amniotic fluid and eat the placenta, an act known as placentophagia.

Djungarian hamsters, noted in the previous chapter for their remarkable paternal care, go by many names. Siberian hamster, Siberian dwarf hamster, Russian winter white dwarf hamster—take your pick. They could also be called delightful. They are small fluff balls of fur, weighing less than two ounces, and come in a variety of colors. They are monogamous, and dads assist in the delivery, help lick off the babies, remove the membrane sac, and consume the placenta. All that licking and cleaning also serves to cover newborns in the scent of the parents. This is pretty standard across many species and appears to promote olfactory recognition of offspring by the parents and of the parents by the offspring.[26]

No one is suggesting you lick your infant off, and while there does seem to be a trend of late to eat the placenta or dry it and grind it up and sprinkle it over food, historically, consuming the placenta is not characteristic of humans. Other animals typically eat the afterbirth to reduce the likelihood of attracting the unwanted attention of a predator that could sniff out the vulnerable mother and newborn. Although the human placenta contains the same hormones, opioids, and volatile chemicals found in other species that may reduce pain and prime parental behavior toward newborns for both mothers *and* fathers, there is no evidence that this is true for humans. Human cultures have historically revered the placenta, *not* eaten it! Despite the current fad and regardless of the claims, there is no empirical support that eating the placenta carries any health benefits. So just . . . don't.

Australian skinks do a little something different. Lizards aren't typically known for their extended parental care, but let me introduce you to the Australian skink, or blue-tongued skink, aptly named for its, you guessed it, blue tongue. For these skinks, the best way to confuse a predator is to have a tail that looks like its head, sort of. Their tail is pretty fat, giving them a bit of an unusual shape. They are also monogamous, and pairs have been known to stay together for upward of twenty years.

Baby blue-tongued skinks stay with their mom and dad for several months after they are born. The babies are viviparous, or live-born, but the kids, *not* the parents, eat the afterbirth. Skink parents, though they don't lactate or nurse their young, still do play a key role in raising their young: They consume less food so they can stay alert, ever watchful for predators—something they should only do for babies that are their own. Thus, even skinks must be able to tell their offspring apart from others. Experiments with mother skinks revealed they do this through touch. They nudge and tongue-flick, and they were even found in one study to prefer to come in physical contact more frequently with a bag that held their offspring than one that held a baby that wasn't theirs.[27] Of course, a reptile's tongue-flicking is also a way of smelling, since it activates the vomeronasal organ, or Jacobson's organ, a specialized organ involved in pheromone detection.

What does this mean for human parents? What we are now coming to understand about humans is that all the "goo" that human infants are covered in, their amniotic fluid, contains many of these same hormones and chemicals (e.g., prolactin) found in other species, including reptiles like skinks, and is the catalyst for bonding between infant and parent through mutual recognition. And remember, this is all separate from that rush of hormones—cortisol, testosterone, prolactin, oxytocin, and vasopressin—that are released in both men and women when their infant is born.

This is about the fragrant fingerprint all of us have, a unique chemical signature mediated by our genes—specifically, our major histocompatibility complex (MHC) genes. This is often generally referred to as pheromones. Aside from their role in mate choice, it is also widely believed that the distinctive scent we all have as a result of these genes also aids in identification of our relatives. One indicator of this is that family members who have had no prior experience with an infant can determine which infant they are related to.

Scents can even shape the way a baby nurses. Human infants exposed to the scent of their own amniotic fluid less than twenty-four hours after birth showed an increased suckling response.[28] We will return to the topic of nursing in Chapter 5, but briefly, the difficulty some infant-mother pairs have with nursing could be alleviated if we would just let newborns get a whiff of home. They may more readily recognize the scent of mom, act calmer, and in the absence of distress, have less "difficulty" nursing straight away. This is just one of the sensory requirements disrupted when newborns are separated from their mothers shortly after birth. Although it may be medically necessary in some cases, it seems more of a convention that can not only impair the parent-infant bond but also lead to difficulties in later life for the child.[29]

Rightly so, many parents these days are insisting on holding their newborn immediately and for an extended period of time.

Touch Me, Clutch Me, Hold Me Close

Since we are on the topic of recognizing one's offspring and the various ways we and other animals accomplish this task, I would be remiss not to mention touch. As I revealed above, many parents of other species lick off their newborns. By licking off a newborn, or other forms of touching, a parent is also covering their newborn with *their* scent. But it may provide an additional function: Physical interaction between a parent and their newborn allows for determining which newborn is yours.

Could this be why human moms have heightened tactile sensitivity after giving birth? If you recall from above, just looking at pictures of human and animal infants enhances the physical coordination of adults, both male and female, and serves to elicit greater caution with a fragile newborn. However, touch serves another purpose. When tested for recognition of their newborns based on touch alone, 65 to 86 percent of mothers were accurate as long as they had spent one or more hours in contact with their infant. With less time than that, the majority failed.[30] This applies, by the way, to fathers, with the same caveat about exposure time. Further, dads reported using physical features such as the size, fattiness, and smoothness of the hands of their infants to identify them.[31]

Surprisingly, there is little information regarding other species on whether parents can recognize their offspring by touch alone, probably owing to the difficulty in isolating that sense for research. What we do know is that skin-to-skin contact is essential to the development of the parent-offspring relationship. It also has the added benefit of raising infant body temperatures better than swaddling in a blanket.[32] Despite this, under even the best of circumstances, extended skin-to-skin contact is not typically practiced in many westernized hospitals. Under cesarean surgical scenarios, it is even less common.

This is particularly true in the case of women undergoing a C-section where physical contact with the mother is often limited even if she is awake. However, it looks like dads can come to the rescue—under both the best- and worst-case scenarios. Infants are normally placed in a cot *next to* their fathers as opposed to directly in skin-to-skin contact. Those babies placed directly against their dad's skin cried less, and were drowsier and calm within sixty minutes of being born, while it took infants placed in a cot next to dad twice as long.[33]

Skin contact was also reported to assist in coordinating the suckling and rooting behavior of newborns, something we know from research on moms. So being in physical contact with either one of your parents results in less distress. Among virtually all other mammals, infants are in physical contact for an extended period of time after birth.

WILD LESSONS

* New parents can learn to recognize their infants by touch with at least one hour of physical contact.
* Fathers can focus on the feel of their newborns' hands.
* Consider requesting to hold your infant rather than have her or him swaddled, as this should increase your baby's core body temperature more effectively.

* Skin-to-skin contact with the mother or father is equally effective at calming and soothing a newborn.
* In cases where a cesarean section is required, skin-to-skin contact with the father is preferable to placing the newborn in a cot.

Why Cry?

Normally, when a newborn comes into this world and is whisked away from its parents, it gives a distress call: It cries. So it seems appropriate to devote a full discussion to crying. And yes, I'm putting my hard hat on and going there for this controversial subject. I've decided to tackle the question: Does letting your offspring cry without responding to them serve a biological or evolutionary (survival) function?

It's not a modern debate. Charles Darwin devoted considerable space in his book *The Expression of the Emotions in Man and Animals* to the subject and starts off the chapter on weeping with the statement, "Infants, when suffering even slight pain, moderate hunger, or discomfort, utter violent and prolonged screams."

In contrast, in the 1940s, Dr. Benjamin Spock, a pediatrician and bestselling author widely known as Dr. Spock, dismissed the idea that babies were suffering when they cried and advocated letting them cry, lest they be unprepared as adults to face the cruel, harsh realities of the real world due to receiving comfort from their loving parents as infants. This position (which Dr. Spock eventually backed away from) was later reinforced by Dr. Richard Ferber, another pediatrician who advocated a similar approach to sleep training, commonly referred to as *Ferberizing*.

Before I weigh in on the matter, let's take a quick look at how other species respond to the cries of their offspring and what's going on in the brain of a crying human newborn. There is substantial evidence that the crying emitted by infants, human and animal alike, is largely under the control of the brain stem and the limbic system. Neuroscientists think of the brain stem as the reptilian part of the brain, not because of its resemblance to reptiles, but because it is considered the old brain, evolutionarily speaking. It is the life support system, regulating functions such as sleep, breathing, and blood pressure. There is definitely not a lot of complex mental gymnastics involved in crying when infants are under a certain age.

Another part of the brain involved in crying is the limbic system. You can think of the limbic system as the emotional switchboard center of the brain. It includes the cingulate gyrus, also called the limbic cortex, which helps regulate emotion and pain. The limbic cortex is like the emotional powerhouse of the brain and is involved in fear and attempting to predict, and therefore avoid, pain.

The limbic system also includes the amygdala, the thalamus, and the hippocampus. Combined, these structures are directing emotional traffic, processing deep social attachment and trust, as well as sending up flare signals to initiate the body's fight-or-flight system. As an infant is responding to its environment, these brain structures are activated, and crying is produced in an effort to receive assistance and stop whatever is causing distress. The only difference between how this mechanism works for infants compared with older children and adults is that, as we get older, we progressively develop the ability to reason through the activation of these brain centers.

The brains of infants alert us to when they are suffering, either physically or emotionally. And they suffer more by the lack of comfort or response given to them by the very individuals charged with comforting and protecting them. This rejection is indelibly imprinted on their growing, developing brains; on the very structures that are responsible for emotional memory, establishment of deep trust, and regulation of emotion.

Given that the oldest and most primal part of the brain is involved in both crying and our sense of fear, crying seems, as Darwin believed, adaptive. From a behavioral perspective, we can weigh the costs and benefits to examine whether this idea holds up. Right away we can see it's vital that the benefits of crying outweigh the costs—because it's noisy and loud. Why does that matter? Crying doesn't just attract the attention of parents; it attracts *everybody's* attention! And while our crying may no longer draw the attention of a hungry predator, the same cannot be said for the white-browed scrubwren, a bird that reveals crying's evolutionary pitfall.

Named for its white "eyebrow," this tiny passerine, or perching bird, is from the coastal region of Australia. The "scrub" in its name comes from the fact that it almost exclusively lives in the underbrush of scrubby areas.

The consequence of this is that the nests are either on or near the ground. For most birds, eggs and chicks are at risk of being eaten. In many bird species the babies chirp and beg with that characteristic gaping beak when their parents are arriving at the nest. But this is not always the case. White-browed scrubwren nestlings cry out whether their parents are at the nest or away. And all this begging attracts the attention of eavesdropping predators.

For scrubwrens, the predator is another bird called a currawong, which looks a little like a raven, but smaller. Parent scrubwrens keep an eye out for these predators while searching for juicy insects to feed their nestlings. Sometimes the nestlings get hungry before the parents have any food to give them and they cry out, begging to be fed. It is not long before such begging draws the unwanted interest of a currawong. Thus crying, begging, or otherwise making noise can be treacherous. If the parents cannot get back to the nest promptly, they do the next best thing: emit an alarm call. The nestlings understand this to mean "Danger!" and quiet down immediately.[34] As a consequence of this enormous cost and risk, scientific research points to the fact that crying is an *honest* signal of need. This distinction is significant because it contradicts the notion that infants cry to manipulate, which implies crying can be a *dishonest* signal.

Being physically separated from their parent(s) is distressing to newborns. And in some species, this aversion toward separation doesn't stop with newborns. Walrus mothers, for example, are "attachment theory" moms. They stay in constant close physical contact with their newborns immediately after birth and continue to do so for two years. When walrus pups have been orphaned or otherwise abandoned, human caretakers take over to be with them twenty-four hours a day, seven days a week, and provide an abundance of hugs.

Attachment theory was put forth by Dr. John Bowlby, a British psychologist, psychiatrist, and certified psychoanalyst who believed Dr. Spock had it all wrong. He insisted that integrating human psychology with animal behavior science (for example, the work of Konrad Lorenz discussed above) provided the evolutionary framework to understand the needs of infants and young children.

Attachment theory in humans mirrors much of what Darwin suggested (and also what countless women of various cultures practiced long before Bowlby's time). When alarmed or in distress, infants are instinctively motivated to get in close proximity to those they trust in anticipation of protection and emotional comfort. The premise is that this will help develop security, emotionally and socially, and as a result, as the infant grows it will be *more* inclined to explore, knowing that it has a safe home base of love and affection.

We see this behavior mirrored in other species. Prairie dogs—whose babies are darling, as I know well from my studies—are also little dinner packages for everything from coyotes to ravens, and even, at least once, a great blue heron. I'll never forget my confusion when I saw a great blue heron that clearly acted as though it was hunting baby prairie dogs in a field of sunflowers from a height of eight thousand feet in Flagstaff, Arizona. I think the prairie dogs were similarly perplexed; they intermittently stared at the heron and alarm-called over the four hours this odd bird was around.

Danger is everywhere for baby prairie dogs. As a consequence, when pups first emerge from the home burrow, they don't stray very far. Mom also stays close. And if the mom isn't around, there is often another adult within reach. As the weeks go on, the pups venture out a little bit further, but still always within close proximity to the home burrow and the fur of a reassuring adult. It is wonderful to watch their confidence grow, aided by a loving embrace when frightened, a kiss periodically to reassure them they are safe, and cuddling together in the darkness of night. Within a few months, these pups are carving their way in the world.

Attachment-style parenting is also common in many nonhuman primates. And like human infants, when separated from their parent(s) or frightened, many nonhuman primate infants cry. They cannot follow, so crying is the only mechanism by which to maintain physical proximity. Crying is highly conserved, meaning we see it preserved across species. It has deep evolutionary roots, evidenced by its presence across virtually all mammal species that carry their infants. Furthermore, not only do we respond to the crying of other mammals, but also the acoustic structure of infant cries in humans is comparable to that of infant cries of nonhuman primates, even though individually we learn the specific cries of our infants.[35]

One such nonhuman primate is the squirrel monkey, noted in Chapter 2 for its long pregnancy and difficult labor, a small primate found in Central and South America, with an additional population established in Puerto Rico. Mother squirrel monkeys also recognize the calls of their infants, and, when separated from their mothers, squirrel monkey infants cry. In response, the moms pick them up. When their mothers pick them up, squirrel monkey babies stop crying. We know that, just like other species, human infants cry when they are physically separated from their mothers or other primary caregivers. We also know that this crying stops when physical contact is reestablished. In humans, cross-cultural evidence reveals that parenting styles that "indulge" newborns and infants by picking them up when they cry results in *less* fussy babies[36] who cry less frequently.

Once, a friend who'd recently become a new mom called me. She was dealing with an overbearing mother-in-law and she wanted some animal behavior science to back her up. She told me she believed in holding babies and picking them up immediately when they cry. Her mother-in-law was horrified and launched into a lecture about how she was outright spoiling her grandchild. She also not-so-subtly implied my friend was on a clear-cut path to ruining her firstborn.

The mother-in-law's approach was relatively medieval, but she clung to it, claiming it had worked with both her children and they weren't any worse for the wear. She let them cry when they were infants until they learned that no one would come when they cried, and so they learned to soothe themselves. She wholeheartedly believed that to do otherwise would result in impossibly spoiled babies, tyrant toddlers, terrible teenagers, and needy adults.

My friend was far from convinced. She needed scientific evidence to present to her mother-in-law and asked me if there was a single other mammal that did not respond to their babies crying. I had to say no. The crying of a baby is truly a universal language. Even a deer will respond to the cry of a human infant.

Yet, when it comes to crying, many people—not just my friend's mother-in-law—particularly in westernized cultures, inject nefarious intentions into the minds of newborns and infants. We accuse them of manipulation. But this isn't even a possibility in the mind of a newborn or infant less than six months old. Instead, when an infant, human or otherwise, enters into this world, it has been born into an uncertain world with no confidence of parental care. It must learn that it is safe to trust that care.

Despite not having access to modern neurobiology, Darwin had it right one hundred years before Dr. Spock and Dr. Ferber, two pediatricians with no obvious training in evolutionary biology, an insignificant grasp of human and animal behavior, and a lack of knowledge of brain processes, who told parents to deliberately ignore their crying infant. And they did it in the most insidious of ways—by doing just what my friend's mother-in-law did: labeling new parents as "bad" parents who would raise unfit, needy offspring if they dared to respond to their newborn's cries.

Had they doled out advice based on what science says, it would have gone something a little more like this: *In order to be proper parents, you must ignore the suffering, pain, and distress of your infant. It is only by doing so that you imprint on your baby's developing emotional brain that you, his/her loving parents, are completely indifferent to his/her needs. That your baby cannot trust you to comfort, soothe, or provide for him/her in his/her hour of need.*

WILD LESSONS

+ Crying is an adaptive, honest signal.
+ Infants age zero to at least six months are crying out of genuine need.
+ Research shows that when parents respond to crying infants, infants *reduce* their overall crying rate.
+ Consider responding 100 percent of the time to crying (when possible) for the first six months. After six months, implement a slow and gradual change in response time. Start with one to two minutes and add additional time slowly.
+ Request that caregivers follow your guidelines.

The reality is humans, like many other primates, have historically carried their infants everywhere, and in many places mothers still do. We are not cheetahs or gazelles, stashing our offspring while we go hunt for food or graze. For both of these species, the babies' silence is a must, lest they become lion food. It is only safe to cry in response to hearing your mom call for you.

This has not been our trajectory. We were not stashed in caves, dens, or long grass—not if our parents wanted us to live. We were born to be carried. We were born to be held. And truth be told, as an adult, when I cry, I still want a hug from someone who cares.

CHAPTER 4

Adjusting to Parenthood

The Physical, Mental, and Social Challenges

My longtime friend Steven, yet another friend from my waitressing days, recently became a dad. We don't talk as often as we used to, but when we do, nowadays there is a common theme. First and foremost, the joy and wonderment he feels watching his son, now six months old, grow and change on an almost daily basis. And secondly, he tells me how everything has changed in his world. He doesn't have enough time. Not enough time alone, not enough time with his partner, not enough time for his business, not enough time to sleep. His son came into this world peacefully but had severe acid reflux for the first four months. That meant that every time Matthew ate, he was in pain. Every time he was laid down to sleep, he was in pain. For little Matthew, who understandably screamed and cried endlessly, it was misery. It was also misery for his first-time parents. Around two months in, Steven and I were talking and he confided, "I think he is trying to kill me." "I assure you he is not," I replied. "He is only going to almost kill you."

Welcome to parenting. The demands are enormous and may bring you to the edge of your physical and psychological limits.

Fortunately, there is a cascade of events that takes place to prepare us for the transition to parenthood. Whether it involves stored energy reserves, hormonal changes, alterations of brain structures, or shifts in social behavior, becoming a parent is a herculean task we share with other species.

Everything Changes

When new parents say, "Everything is different once you have kids!" they may be talking about time, or sleep, or resources, but there are other, unseen changes that happen first, even before their children are born. That's right: We are kicking this chapter off with hormones, those pesky chemicals that can wreak havoc in our lives when they are out of whack, but are also the very same ones we quite literally cannot live without.

As we saw in Chapter 2, in the vast majority of species, females are the ones who become pregnant. As women, we each experience substantial hormonal shifts over the course of every estrus cycle. For many of us, myself included, these normal fluctuations can result in some odd thoughts, feelings, and behaviors. I like to think of myself as a rational, put-together person. But a few days out of the month, I will weep at the sight of a puppy, cling uncomfortably to a friend as I say goodbye (even though I will see her again within twenty-four hours), and be plagued by indecisiveness over my entire life plan at around three in the morning.

It should come as no surprise then that with pregnancy there is an even more dramatic increase in the levels and types of hormones coursing through our bodies. These aren't willy-nilly rises and falls of hormones, but rather precisely timed shifts to gear us up to be parents. But is it just us? Is it just females? And are we the only species that gets hormonally primed for parenthood?

To figure this out, let's look at three particular hormones: cortisol, prolactin, and oxytocin. These are some of the less-discussed and therefore more unfamiliar hormones, but they play a huge role in the onset and maintenance of parental behavior.

Cortisol may not seem like it should be an active player in triggering parental behavior. After all, some may know it as the stress hormone that increases belly fat when overproduced. Cortisol is one of several glucocorticoid, or steroid, hormones that modulate inflammation and regulate glucose in almost every single vertebrate species and all mammals. It is produced by the adrenal cortex and is released when we experience stress. All types of situations can activate the release of cortisol, including being chased by a lion, food shortages, injury, or too many deadlines. It is the chronic release of cortisol in response to *constant* stress that is problematic for humans and animals alike.

We can see this in gray-cheeked mangabeys living in the mangabey equivalent of New York City. Gray-cheeked mangabeys are large, somewhat longhaired primates, with lovely faces surrounded by long wisps of hair. They are found in various parts of Africa, including the lush forests of Uganda. Normally they prefer unlogged, undisturbed forests, but as humans have encroached into more remote regions, some mangabeys find themselves in areas that have a high human presence. As a consequence, those mangabeys have chronically higher levels of cortisol in their urine and it negatively impacts their health.[1]

We know that chronically elevated cortisol can disrupt reproduction, immune response, growth, and even neurological function, so what the heck is it doing rising just before birth? True, pregnancy is stressful, and there is a separate rise in cortisol levels early in pregnancy, when the release of cortisol stimulates other hormones necessary to set the clock on the duration of gestation and the timing of birth. But like eating a power pellet in the game of Pac-Man, pregnant women also produce a protein that sticks to all the cortisol, removing it from circulation and canceling its negative effects. That is, until just before the baby is born. This is a delicate balance, because the consequences can be pretty severe if too much of this hormone is produced. If you are one seriously stressed-out mama, then the high levels normally produced during this time are compounded with even more cortisol. This can set up a chain reaction, increasing the risk of miscarriage, pregnancy-related hypertension, premature birth, and abnormal child development.[2]

That could be what happened to Caroline, from Chapter 2 (the woman with the liver problem). She delivered her daughter prematurely by cesarean section. Plagued by anxiety and an obsessive-compulsive disorder, Caroline worried about everything. All that anxiety made Caroline a very high-strung mom-to-be and potentially contributed to the adverse health outcomes she experienced during her pregnancy. Although there are no obvious signs of developmental or cognitive issues in her delightful daughter, sometimes the consequences of excessive exposure to stress hormones during gestation don't manifest until much later. Even if there are no physical problems that develop, excessive stress can cause things to go wrong in the bonding between a mother and her offspring, resulting in maladaptive parenting and neglect. Though we talk more about maladaptive parenting behavior and abuse in Chapter 8, it is worth mentioning here that, collectively, we as a species have created an environment for ourselves that causes us to experience abnormal and chronic levels of stress, which is affecting our children and our ability to parent. The consequences of this have enormous social and economic ramifications and are worthy of our attention.

Given such risks, why do levels of this "red alert" hormone surge just before giving birth and stay elevated for a while? One contribution is from the soon-to-be-born baby. Just before an infant is born, he or she initiates a flood of cortisol to be released from the placenta as a final push for growth, neurological regulation of emotions, cognitive development, and lung maturation.[3]

But there is another reason. The interesting part of the cortisol story is that levels remain elevated even after the baby is born, implying an increase is simultaneously happening in the mother. This rise influences the ratio of the sex steroid hormones estrogen and progesterone, which in turn are linked with maternal behavior. Research on some nonhuman primates is beginning to shed light on these relationships.

Living in the Horn of Africa, hamadryas baboons were considered sacred by Egyptian deities such as Thoth, the god of knowledge. These baboons live in a multilevel society run by males, but females do the bulk of the parenting. When looking at the relationship between cortisol and the interactions new baboon mothers had with their infants, it was discovered that those mothers who had higher levels just before their infant was born exhibited stronger maternal behavior. This completely changed if the cortisol levels remained very high for too long after an infant was born. In that case, it reflected higher stress levels in mothers, which lead them to interact *less* with their infants.[4] This means that the benefits of high cortisol levels are strongly linked to how *long* they remain elevated in the mother after an infant is born.

How does this play out in new human moms? In very much the same way. As I mentioned, just before birth the amount of cortisol circulating through a pregnant woman's body is extremely high. For about one week after a baby is born, new moms who have higher amounts of cortisol physically interact more with their infants by holding them more. They also find the smell of their babies extra delightful, they do a better job of recognizing them based on smell alone, they have a stronger sense of empathy when their infants cry, and, finally, they generally feel more well-adjusted.[5] It is likely that this physiological response is needed to keep a mother on alert and all her senses active and heightened. And just like in the hamadryas baboons, this effect is temporary.

Human males don't get pregnant, but that doesn't mean that they won't experience hormonal shifts that gear them up to be fathers. When it comes to cortisol, their levels don't rise until the very end of their partners' pregnancies. And, just like in women, it is thought that this short-term peak in cortisol serves the same function: to facilitate interaction and positive paternal behaviors.[6]

Another fascinating hormonal change taking place involves prolactin. This hormone is most well known for its role in maternal behavior and milk production in mammals. Prolactin is involved in over three hundred biological functions, and what you may not know is that it is often referred to as the hormone of "fatherly love."

I love mourning doves. I love the way they coo and gurgle. They are elegant and bumble around all at the same time. They are a migratory bird belonging to the columbiform group that also includes pigeons. They mate for life—until death do doves part—and the pair, despite going their separate ways during migration, manages to come back together during the breeding season. Both parents incubate the eggs and feed their chicks. An unusual feature of this group of birds is that they produce crop milk for their nestlings. This isn't the same liquid gold that mammals make, but it surpasses mammal milk in its protein and fat content. That's pretty remarkable since it is secreted from the lining of the bird's crop, a muscular esophagus-like pouch near the throat used to store surplus food.

This substance is critically important, especially in the first four to five days post-hatching. As with lactation in mammals, prolactin is crucial to the production of this rich and nutritious food source.[7] The kicker is that it is equally important in males as it is in females. I must confess that even though for three years I had the same pair of doves nesting on my patio, never once did I realize that either of them produced milk for their chicks!

But what about other species that don't produce crop milk? Does prolactin still matter? Yes, it does. It makes the difference between being an involved father versus an inferior one. The blue-headed vireo is a striking songbird found in North and Central America. Its songs are simple and repetitive, the Justin Bieber of songbirds if you will. They are primarily insectivorous, eating everything from beetles to bees. Right off the bat, males show off what great dads and providers they can be. A male builds a courtship nest to display his nest-building prowess to a female and, if she is suitably impressed, she will mate with him. Then they work together to build the real nest for their future family.

Clutch size in this species is anywhere from three to five eggs, and the dad is heavily involved in sitting on the eggs, upkeep of the nest, and feeding the chicks—as long as his prolactin levels are just right. A tiny bit lower and he would be contributing less to the incubation parental duties, as happens in his cousin the red-eyed vireo.[8] Male red-eyed vireos contribute in all other ways, including the feeding of the chicks, but they do not normally incubate the eggs.

By now you might be wondering if the role prolactin plays in fatherhood is just for the birds. Of course not! Titi monkey dads are model fathers. This small monkey has large eyes because of its nocturnal lifestyle. A male and female live together in a family unit that may include older children. Titi monkey males take their parenting duties very seriously. The male carries around the new infant the vast majority of the time, only allowing the mom to nurse the infant. He is so involved that the infant may bond more strongly with him and experience greater distress when separated from him versus the mother. When it comes to prolactin in these superdads, research reveals that titi monkey fathers have higher levels of circulating prolactin than their adult sons who have not yet fathered children.[9]

Obviously, like birds and monkeys, human males do not have the benefit of direct pregnancy-induced increases in prolactin, though men that experience couvade syndrome, mentioned in Chapter 2, do have higher prolactin levels than men with few to no pregnancy-type symptoms.[10] But for the rest, there is a positive feedback loop between contact with their infants and prolactin, similar to what is seen in other species. When dads play and interact with their infants, there is a corresponding increase in their baseline prolactin levels as well as increased concern when their new babies cry.[11] Don't worry, men, you won't start lactating because of this, though you might if you let someone suck on your nipples too much—or if you are a Dayak fruit bat.

Let me explain. In the case of human males, there has been some evidence of lactation, but it is rare. Repeated stimulation of the nipples will cause a rise in prolactin levels and induce lactation in females that aren't pregnant as well as in males. But in the case of the Dayak fruit bat, males seem to spontaneously and naturally lactate.

In 1992 in the Krau Game Reserve in Malaysia, a couple of scientists observed something odd. While mist-netting bats (a mist net is a special type of net used by scientists to catch birds and bats) these scientists caught ten mature male Dayak fruit bats. Dayak fruit bats are pretty rare and not a lot is known about them, but much to the researchers' surprise, they discovered that all ten had fully functional, milk-expressing mammary glands.[12] This is the only known mammal where both females and males lactate.

Another hormone that is in the news frequently is oxytocin. It is referred to as the bonding or feel-good hormone, but this is a bit of a misrepresentation. We should really refer to it as the *feel* hormone. That's because oxytocin strengthens all kinds of social interactions, positive *and* negative ones. For instance, if you are interacting with someone you distrust, oxtyocin can be released and amplify your negative feelings toward that person.

However, oxytocin is heavily involved in the positive bonds between partners and between parents and their newborn infants. For quite some time, it has been clear that oxytocin increases uterine contractions during labor and triggers moms to behave in a nurturing and loving manner. If you give a female rat that has no pups enough oxytocin, she will build a nest and go get someone else's pups and carry them back to her nest! When it comes to human moms, oxytocin levels are linked to how much new moms touch their babies, look at their babies, and even how they feel about their babies.[13] And it extends past the infant stage and influences how receptive mothers are toward their toddlers.[14]

What does this mean for us in this stressful, jam-packed, busy life we have created for ourselves? Some of us are missing the opportunity to enhance and solidify the bonds and relationships with our kids. The research tells us that frequent physical contact and positive interactions increases oxytocin levels. And it's not just about moms. Common marmoset parenting behavior gives us some insight into oxytocin-driven affection, attentiveness, and caring behavior for dads as well. These small primates, discussed in the previous chapter, have two fluffy white poofs of hair adorning the sides of their heads, making them look like they stuck their fingers in an electrical socket. As they gallop along branches, these tiny primates feed on insects, fruit, nectar, tree sap, and maybe a frog or two. After chowing down, they usually take a nap. Because they are so small, some people think they should make good pets. It turns out, they do not, as an old friend of mine, Trevor, discovered. He used to own a marmoset and he also used to be a male model. One day this tiny, less-than-one-pound primate took a bite out of his nose. He is now on the other side of the camera.

Marmosets live in a tight-knit family group composed of the mother, father, and offspring. What is cool about them is that they are part of a select group of primates that always produce nonidentical twins. Although male common marmosets cannot nurse their offspring, they do just about everything else, including carrying, protecting, grooming, playing with, and, when the kids are old enough to start learning the marmoset diet, transferring food to them. The dads may do this for many months, gradually reducing the frequency of feedings and their tolerance for begging by their kids. However, when marmoset dads were given more oxytocin, they became willing to continue sharing food, even with offspring that were at an age where dad would normally stop.[15]

Some dads feel apprehension about holding their newborn infants. Craig was like that. I met Craig when I applied for a job in South Africa at a game farm he managed. I didn't get the job, but we became fast friends online, met in person several months later, and are still close today. He is a tough, burly bush biologist, but when it came to holding his daughter, he was so nervous! He didn't want to hold her the "wrong" way and was convinced that she would break. His wife, however, was confident in his abilities and encouraged him to hold her frequently. It didn't take long before he was behaving like a possessive marmoset father, wanting to hold her, carry her about, and play with her constantly. He can thank oxytocin for this shift from anxiety to confidence, from reluctance to enthusiasm. Well, also his wife, but we know that as human dads physically interact with their infants— gazing at them, talking to them, and playing with them—oxytocin levels rise, driving them to interact even more.[16]

One way, then, to increase oxytocin levels is to interact physically with your infants and toddlers—and to keep doing it even as your kids get older. Oxytocin is on a positive feedback loop. The more you interact, the higher levels rise, driving you to interact. This means putting the electronics down, getting outside, and playing together. It means hugging, kissing, and being affectionate with your children constantly. Even brushing their hair, a form of grooming, is a simple act that will promote closeness. Read a book together, play a board game, hold their hand.

Given the significant hormonal fluctuations and influences governing both mothers and fathers during pregnancy, after birth, and into the beginning days of life with a new infant, it seems timely to mention postpartum depression. For many women (and men), bringing new life into the world and into their families fills them with a sense of joy, eagerness, and anticipation. But sometimes those hormones go awry and postpartum depression happens. There is a spectrum of severity, but we know that roughly 15 to 20 percent of women experience some degree of acute postpartum depression, while the mild postpartum "blues" affects 40 to 80 percent of women. The reason for such wide variation has more to do with defining it than what it means for any single woman to experience it. However, most episodes start shortly after delivery. Although inconclusive, the dramatic changes in hormones associated with pregnancy, and subsequent withdrawal of those hormones, which decline rapidly after the first week, may be a catalyst for postpartum depression in some women.[17]

As a society we are improving our understanding and compassion for women who experience postpartum depression, but by and large I think many new mothers are under the impression that they are only supposed to feel ecstatic and are failing as moms if they experience or suffer from other emotions. As a biologist, it is obvious to me that hormonal imbalances and rapid shifts can dramatically affect new mothers (and anyone else, by the way). Another problem is that, although many moms expect to feel fatigued, irritable, and unable to sleep, women with postpartum depression also experience these symptoms, so it can be challenging for moms, family members, and doctors to discern when the problems have surpassed common postpartum experiences. One key is that moms with ordinary post-pregnancy symptoms still look at their babies and feel joy and affection.

I remember my friend Jackie went through this after her son was born. At the time of her son's birth, I had known her for only a short time. Still, it was immediately apparent that she was put together, a hard worker, and joyful and happy about the arrival of her son. Within two weeks of his birth, however, she began to experience extreme anxiety and apprehension about being left alone with him. She was inexplicably frightened that she would do something wrong or hurt him. Her husband, rather than offering support, demonstrated his lack of understanding by criticizing her as a new mother. She was devastated, and the lack of help and understanding prolonged her suffering.

Aside from affecting hormones, the delivery of a baby can disrupt the expression of certain genes that alter brain activity in key regions through neurotransmitters, hormone receptors, and genes that influence anxiety-related behavior. What this means is that there are very clear biological explanations for this condition. Given that, as I have already described, we are not the only species to undergo massive hormonal fluctuations during pregnancy, delivery, and postpartum, is there any evidence for postpartum depression in other animals?

As challenging as it is to determine if postpartum depression is occurring in human moms, the odds of assessing if it's the cause of neglect by some wild animal moms are slim. However, we do know that some animal mothers inexplicably reject their offspring or may not be as attentive or affectionate as they should be. And research on rats given hormone treatments simulating the rapid rise and fall of hormones associated with pregnancy and birth revealed that they had symptoms of depression when the hormone treatments were stopped. Incidentally, depression in female rats was measured in terms of reduced movement during different tasks.[18]

Whether marmosets or humans, individuals—both mothers and fathers—vary in hormone levels, which, in turn, affects how attentive, affectionate, and caring parents are toward their newborns. These early interactions and experiences set the stage for the relationship parents develop with their children over the long term and for their health and well-being through development. As a result, their importance should not be ignored or trivialized. Nor should we ignore or trivialize the struggles that many new parents face in general, including when these chemicals become imbalanced.

WILD LESSONS

* A rise in cortisol levels is necessary and beneficial at various stages of pregnancy and beyond, but it has its limits. Too much stress has cascading negative effects on moms and infants alike.
* Prolactin isn't just for moms and birds. Human males may not nurse like Dayak fruit bats and certainly can't produce crop milk, but prolactin is behind the scenes bonding dad and baby.
* Oxytocin is front and center in both moms and dads in influencing affection.
* Never underestimate the power of hormones in driving what you think, feel, and do. When you feel off in your attentiveness, affection, and joy toward your newborn, imbalances may very well be the culprit. Seek immediate assistance, and if your doctor won't listen, find another doctor!

continues . . .

Have I Lost My Other Mind?

It is a given that all the hormonal changes discussed also affect the soon-to-be parent brain. But simply becoming a parent transforms the brain, changing different structures and connections to meet the physical and emotional demands of raising a child. And these changes happen whether you are a parent to a biological child or an adoptive child. These changes in neural circuitry are integral to supporting parental behavior in humans and other animals. Essentially, the brain must be rewired and reorganized to become a parent!

Across all mammals, one crucial brain region is the part of the hypothalamus called the mPOA. The hypothalamus is about the size of an almond, squished between the thalamus and pituitary gland, and has an intimate relationship with the hormones released by the pituitary gland, including all of the ones discussed above.

The mPOA provides sensory feedback (visual, olfactory, temperature, etc.) in combination with pregnancy-related hormones to change the structure and activity of neurons. These neurons then interact with other parts of the brain to ensure that mothers are receptive to their new infant.[19] You probably thought it was just automatic. Oh wait, it is! It's biology!

Remember Julie from Chapter 1? The one who obsessively relayed every detail of the events surrounding bringing sweet Sam to lunch? She's not alone. New parents seem to be obsessed with their babies, talking incessantly about them, laser-focused on everything from how often they eat to those seemingly endless poop discussions. This obsession has a name: maternal preoccupation, or maternal constellation.[20] Modern neuroscience is shedding light on what parts of the brain are at work here, what drives moms and dads to be engrossed with every aspect of their infants, including how "flawless" they are. For moms especially, the intrusive thoughts are centered on fears and anxieties about the safety and cleanliness of their infants. From an evolutionary perspective, it makes absolute perfect sense for selection to have favored this. If your baby isn't safe and if your baby isn't clean, it's very survival could be at stake. Virtually all mammal moms in the animal kingdom—and dads, when they are involved—lick, clean, or otherwise groom their new babies, and they do so almost obsessively.

My friend Paul felt an altogether different type of anxiety. Paul and I met during graduate school and about a year in he had his first child, a daughter. He is such a steady, thoughtful, calm person that I was very surprised when he revealed how he fretted over his daughter. In his case, he was not so concerned about his daughter's cleanliness, but instead he had a heightened sense of danger—to him, danger lurked *everywhere*. His wife was in school at the time, and so, after overcoming his initial apprehension about holding his child, he happily took over a large portion of the parenting duties. The building where his wife went to school had incredibly steep staircases that were structured like square towers. There were three segments of stairs and a landing per floor, so the stairs were a squared spiral around a massive, open, empty chasm—and, of course, his wife's studio (she was an art student) was on the top floor. He shared that whenever he would use these stairs while carrying his daughter—and he had to, as there was no elevator—his mind would flood with terrible images of her falling out of his arms and plunging down to splat at the bottom.

His imagination was vivid and absolutely uncontrollable, so he found that the best way to cope was to let the nightmares play out. This was unsettling and disturbing, to say the least, but then, perhaps because he didn't resist them, he began noticing similar imaginings in other places of danger. Ultimately these disturbing thoughts brought him comfort, as he explained to me: "She was so important to me, and my role as her protector was so crucial, that my mind was sending me constant reminders of the horror that could unfold if I wasn't careful. I always hugged the wall when I carried her on those stairs, and the visions of disaster can be credited for that behavior."

Like Paul, in countless other situations parents' minds envision disaster befalling their children, precisely so that disaster won't happen. This kind of imagining is normal and a consequence of the structural reorganization of the brain that has evolved to vividly identify and model risk to offspring.

Aside from the mPOA, the ventral tegmental area, or VTA (also linked to romantic love), becomes highly activated when men and women become parents. This area of the brain is part of the meso-limbic pathway, also referred to as the reward pathway, located in the midbrain. The VTA has a high density of dopamine neurons, and it is one of several pathways that send dopamine out to other areas of the brain. Dopamine is the pleasure hormone, and the rush of this neurotransmitter makes people feel happy. This action creates a very strong motivation—resulting in a reward feedback system between parent and child, compelling parents to take care of and feel positively toward their infants[21] and bringing happiness to mothers and fathers when they have contact with their infants. It's quite literally a natural high!

And when it comes to protectiveness, look out! Most moms (and dads) will aggressively and viciously protect their offspring against a threat, even one they might not normally fear themselves. We often say that a mother will defend and protect her offspring like a "mama bear." Perhaps that is because the fastest way to get attacked by an actual bear is to come between her and her cubs. It doesn't matter if it's an accident. Brown bears, black bears, and polar bears are not exceptional in protecting their cubs from danger, but there is something special about brown bears in Sweden. There, mother bears do not avoid areas with people, but instead actively go near human communities. It sounds like a pretty risky thing to do, and mind you, I mean for the bears, not the people.

What in Sweden is making these mother bears bring their cubs in close proximity to an enemy? Male brown bears. That's right, the threat that these adult males pose to cubs is so severe that a mother bear will overcome her trepidation of being in close proximity to humans in order to shield her cubs. Essentially, she's driven by the sense that "the enemy of my enemy is my friend." In the case of these mother brown bears, staying 1,500 feet closer than other bears to humans makes the difference between life and death for cubs.[22]

The ability to overcome a fear, to face a dangerous threat head-on to protect your children, is common to many animals, including humans. Although there are many different, interconnected parts of the brain that shift, grow, or otherwise change to facilitate aggressive, valiant defense and protection of offspring by both mothers and fathers, I want to focus on the two almond-shaped parts of the brain, one on each side, called the amygdala. (As a side note, I am beginning to wonder why so many parts of the brain are typically compared to nuts, especially almonds?) The amygdala is considered part of the limbic system and has its fingers in rapid decision-making, memory, and emotion. It is a powerful force in the fight-flight-freeze system.

A testament to the power of this system was a confrontation between a human mother and a mountain lion. What stood between them was the woman's five-year-old son—or rather, the mouth of the mountain lion! I've only seen a mountain lion once in the wild, and I can assure you that approaching it or wrestling anything out of its mouth would not be something I would normally consider doing, and I'm quite certain this mother never thought she would, either. According to the mother, as her children were playing outside one evening, she suddenly heard a scream. When she went outside she discovered that a mountain lion was on top of her son. Without even blinking, which is exactly how fast the decision-making process happens in the amygdala, she rushed at the animal to find that her son's head was in its mouth. We've heard the stories of people who have the strength to lift cars off of someone—well, this mother pried open the jaws of this mountain lion to get her son's head out of its mouth.

That's not the only example of a stark change in brain chemistry caused by parenting. Yet another striking change to the brain (and other parts of the body) that might seem quite unexpected is that an infant's cells are released into the bloodstream of its mother and will be found circulating in her body for life. This means it's literally true that your children are part of you for life. A chimera in the mythological sense refers to mystical creatures that are hybrids of different parts of animals put together. In *The Iliad*, the Chimera was the fire-breathing monster with the body and head of a lion, a goat head coming out of his back, the udders of a goat, and a serpent for a tail. What does this have to do with being a mother? When it was discovered that the mother's body contains the genetic material of her child for life, scientists called this phenomenon a "microchimera."

Here's what happens. During pregnancy, stem cells from the fetus cross through the placenta and enter the mother-to-be's bloodstream. Straight out of a science-fiction movie, these cells, which have the advantage of turning into any kind of tissue, float around the mother's body and, depending on where they settle, start to grow into that specific type of tissue. The mother's immune system seeks out and destroys the cells when it finds them, except for those that have already embedded themselves into tissue. And this happens with every baby that a woman has. Over time mothers accumulate cells from all of their children. As with many interesting scientific discoveries, this phenomenon was detected by accident when male DNA was found circulating in a woman's body.

There is still much to discover about why this happens, but it is speculated that the cells from the fetus settle in areas that benefit *itself* the most, such as breast issue to affect the lactation, the thyroid to modify metabolism, and the brain to influence the circuits involved in maternal behavior.[23] We know this happens in other species—like rats, who have served as model organisms for studying many of these processes, but my guess is that this is common to all placental mammals.

And finally, rounding out some of the physical changes that happen to new parents, we have cognition and memory. There are tons of jokes surrounding what we might call "momnesia" or "mommy brain." Even Leila, when she was sharing her story about being pregnant and having loose hip joints, laughed when I asked her about any changes in her memory. She chuckled because she meant to tell me about it, but she forgot. So what gives? Is it true that moms and dads become more forgetful? Or do they have cognitive gains in some areas but have to compensate by giving up a little bit on the short-term memory side?

Prairie voles are the "ideal" family unit. Males and females form a close bond, and both are virtuously monogamous unless they get drunk. Although many animals do enjoy throwing back some fermented fruit and can take things too far and become inebriated, the only time prairie voles drink alcohol is when scientists, who want to see if they remain faithful when under the influence, provide it to them. When it comes to raising the kids, both mom and dad voles are super-involved parents.

These marvelous, short-lived, tiny rodents make their home in the grasslands of North America. Recent research has revealed that they are empathetic, or sympathetic, to the distress of their mates. When a prairie vole's partner is stressed, the other gives it the equivalent of a massage to help calm it down.[24] Parenting can be stressful for both moms and dads, and by being sensitive to when your partner is approaching his or her limit, stepping in and offering a hand (or even a massage!) can do wonders.

When it comes to parenting behavior, studies have found that male prairie voles with even the briefest contact with pups showed enhanced cell growth in the dentate gyrus, a part of the hippocampus, that is linked to memory and navigation.[25] This is one of the few areas of the brain that is capable of growing new neurons, a process called neurogenesis. We usually grow new neurons daily, but certain events and experiences can trigger rapid growth. Neurogenesis is believed to be important in the creation of new memories. Normally, nonparents show enrichment of the hippocampus when exposed to new situations or environments, yet you see the opposite effect in biological parents. What's going on here?

It all comes down to those pesky hormones. For mothers the phenomenon is more extreme. Reduced levels of estrogen and increased levels of cortisol, the stress hormone, render pregnant women and new moms unable to produce new neurons. And they actually *lose* hippocampal volume. Overall, pregnancy negatively influences a woman's short-term working memory, and it continues, getting slightly worse just after the baby is born. The ability to recall information is also reduced. And finally, the speed at which the mother can process information declines. Those deficits can be attributed to what's happening in the hippocampus,[26] which might be why Alma (who, as discussed in Chapter 2, developed a lifelong aversion to baby corn) couldn't remember what she couldn't remember when she had her daughter twenty-two years ago, or Leila couldn't remember to tell me that she couldn't remember things that well anymore!

Even though dads don't experience the same dramatic shifts in hormones that women do, you will recall that they do experience rising cortisol levels. Therefore, their neuronal growth is also suspended. Fortunately, for both males and females, this effect is temporary.[27]

Forgetfulness can have devastating consequences, particularly for activities you don't do frequently or every day. Like taking your infant to daycare. It's possible that some of those instances where, for example, a parent who rarely takes his or her infant to daycare and ends up leaving the baby in the car are a consequence of the memory and cognitive deficits that happen to new parents. This can lead to horribly tragic consequences. And it seems easy to think that loving parents could *never* forget where there child is, but we know that isn't true. Much to the child's dismay, many a parent has forgotten to pick his or her older kid up from school. The reality is that, on a day-to-day basis, there are a million things you could forget—and you probably do! Misplace your shoes? Drop the kids off at school without their lunches? What appointment?

It would be easy to say: "Just be more organized." But *how*? One approach, if you have a partner, is to divide and conquer. Write things down on a big board in the house, send text alerts (repeatedly), and especially look out for each other when something in the routine changes, as that will almost guarantee forgetting.

But it isn't all bad news for your brain. The prefrontal cortex (PFC)—the part of the brain that is like a giant cognitive processing center that plays a role in planning, your personality, social interactions, regulation of your emotions, and decision-making skills—is also reorganized once you become a parent. Unlike in the hippocampus, there is an increase in the density of neurons in the PFC. Once again, this happens in both mothers and fathers. Remember those lovely marmosets mentioned earlier? Research shows that new and experienced fathers demonstrate enhanced activity in this area of the brain.[28] Unfortunately, this effect is also temporary and fades as children get older.

WILD LESSONS

* Mothers, well at least human mothers, don't really have eyes in the back of their heads, but many of the changes that happen in the brain allow them to be more focused and attentive to everything that is happening with their children.
* You will inevitably find yourself imagining possible impending doom for your children. Perhaps the only way to not become consumed by it is to accept it and reframe your worst imaginings by understanding and appreciating how evolution has shaped the circuitry of the brain to help you protect your children.

continues . . .

- And when it comes to maternal protectiveness, don't come between a mother and her cubs—be they bears or people. And if you're a mountain lion, there is at least one mother you better stay away from!
- If you've ever felt like your kids are part of you, they literally are.
- There is an interesting contradiction when it comes to your brain. On the one hand, new neurons grow and improve some cognitive functions, while simultaneously your memory suffers!
- Chart daily parenting duties. Send reminders and text alerts, especially if you swap tasks with your partner.

The Parent Club

As you may recall from the beginning of this book, Julie lamented about how she did not even have time to eat, shower, or clip her toenails since having Sam. Cichlid fish can relate because, like the gastric-brooding frog we talked about in Chapter 2, many species in this group mouth-brood as well. Though in the cichlid's case, any food a parent takes in is gobbled right up by the young, and the mom or dad—there is one species where the dad alone handles this—is left starving until the children are old enough to feed themselves. I know, I know, starving is extreme, and human parents find time to eat. But many parents struggle to find time just for the basics.

It's not just a matter of not eating properly. For parents, "me time" of any kind goes out the window, especially when offspring are young and completely dependent. Part of "me time" includes the wardrobe—at least for Julie. She confided in me that a few months after having Sam, she looked in her closet one day and wondered who on earth had worn those clothes. When they have a baby, wild capped langur moms have to give up doing as much for themselves as they normally would. Unfortunately, when I was in Nepal I didn't catch a glimpse of these Old World monkeys. Their range also extends into Bangladesh, China, India, and a place I've always wanted to go, Bhutan. They live in groups that include only one male, and females are fairly amicable with one another. Among other ways that "me time" is lost, new langur moms lose about 32 percent of their time to eat when they have a newborn.[29]

For human mothers, aside from personal wardrobe and hygiene and all the other individual stuff you don't have time for, you have a suite of relationship adjustments to contend with. If you are co-parenting with a spouse or significant other, the demands on your relationship will reach new heights. If you aren't able to figure out how to operate as a cooperative unit and team before the baby arrives, in light of everything we've already covered, your chances of doing so *after* will be remote.

A lot of couples have trouble divvying up the household chores amicably, and with a new baby, the pressure intensifies. We know there are genetic, environmental, and social influences on the degree to which individuals (and cultures) around the world are monogamous and demonstrate equal biparental care, yet it is evident from all the physical changes that *both* mothers and fathers experience when their infants arrive that shared parenting is part of being human. Thus, let's dispel the notion that mothers are the most important parent, that fathers aren't crucial to the development of our children, or that a male's role is simply to go to work and support the family.

As we see in all species, some moms and dads are better than others. So how can we approach shared parenting and negotiate the different degrees to which each parent contributes to care to ensure healthy, successful offspring? And that's assuming said parents haven't lost each other or their minds along the way. Perhaps we can learn from long-tailed bushtits, who have developed a successful strategy. This small passerine, or perching bird, looks as though it might fall off a branch at any moment because its body is disproportionately rotund and tiny compared to the long tail it is sporting. This bird can be found across Europe and Asia and, when they are not breeding, live in a moderately sized family flock.

During the breeding season, the adults split off and form pairs. As we have seen in other birds, both the mom and the dad help raise the kids.

Much like us, long-tailed bushtits face a dilemma. Each parent would prefer if the other parent did most of the heavy lifting. One way to solve this quandary is to only give as much as your partner does. This can also be called tit for tat, or as we refer to it in human relationships, keeping score. Of course, if no one gives, the offspring don't make it. If no one gives enough, some of the offspring may die. What happens among long-tailed bushtits? Each parent arrives at the nest at the same time, monitors the other, and watches to make sure they are providing food before they provide food. Even though they take turns, someone has to go first! I wonder if they even take turns in who goes first? Nevertheless, this keeping watch and keeping score means they systematically and synchronously feed their chicks. The chicks whose parents do this successfully get fed more and survive at a higher rate. As an added bonus, because of this watching over the other parent, both parents are at the nest together and gone together, attracting less attention to activity at the nest, which means fewer babies get eaten by predators.[30]

Not surprisingly, the same kind of harmony holds true for human parents that split feeding duties for a new infant. Sleepless nights are problematic for new parents, and those who take turns make sure no single parent gets too exhausted. That leaves them more energy to pay attention to their relationship as well, something that many new parents think necessarily gets tossed out the window when children arrive on the scene. Although coordinating parenting and other duties is part of maintaining a solid partnership, neglecting the partnership and failing to set aside couple time can prove disastrous.

In many human partnerships a common complaint is no time for sex. It is a myth that other animals only have sex to make babies. Even after the babies arrive, many species still engage in ritualized courtship. For cockatiels, divorce might be imminent if one partner isn't as receptive to the affections of the other. This is one of the principal fears experienced by human soon-to-be fathers, the demise of the husband-wife relationship.[31]

One thing that would help give parents time to adjust is higher standards for maternity and paternity leave. The United States lags behind all developed countries that provide substantial early transitional support and up to sixteen months *paid* (100 percent) parental leave. There is a substantial disconnect between how we say we value families in the United States and how we actually treat families. The lack of support for new parents has consequences for society as a whole, and other countries seem to recognize this and have taken steps to assist parents as they transition.

But even if you've got this parenting tag team going smoothly, other relationships see change and strain, predominantly your social relationships. Barring lifelong friendships, some, if not most, other friendships may fall to the wayside. While new parents withdraw from friends, they tend to increase contact with their family. The exception to this is if their friends also have kids. In that case, those friends are now full-fledged members of the parenting club and invited into the inner circle. My friendship with Julie, whose son is now almost two, has faded over time, and now I only catch glimpses of her life posted on Facebook, where I can see that most of her current friendships are with other mothers.

This isn't so different from the phenomenon of "calving grounds" in wildebeest. As you may recall from Chapter 2, wildebeest mothers only like to hang out with other wildebeest mothers. These groups of moms stay together until their young begin striking out on their own. One of the reasons they do this is because of the safety in numbers. By flooding the Ngorongoro Crater with calves and sticking together, the risk of being eaten by a predator goes way down.[32] We may not have lions on our heels, but Leila put it the best: "Once you become a parent, it is like you suddenly become part of this exclusive club, this network of other parents that get what you are experiencing and are there to help."

But here's the thing: Sometimes your nonparent friends can help, and they want to be a part of your life, too. For the wild capped langurs mentioned above, allomaternal care, which means care by someone other than the mother, is fairly common, and oftentimes it is inexperienced females who want to get their hands on a baby and help. Like many moms, wild capped langurs are reluctant to turn over their babies to other females, especially ones who don't have kids of their own, but the moms who do accept help often reap the benefit of having more "me time," which in their case can mean spending more time eating.

The ideal setting for humans is a close-knit tribal setting with an extended tribal family that includes nonrelatives. Town and city living has created a new set of hurdles for parents. Create your tribe and ask for help. Because new parents are often strapped with work and supervision, it can be difficult to accomplish basic things like cooking and cleaning. If you are reluctant to let a friend babysit, there are other real and measurable ways to receive help. Ask friends to mow the lawn, do the grocery shopping, or prepare a meal. This is what black-faced grosbeaks do.

As part of the family of birds that includes cardinals, these exquisite neotropical birds can be found throughout Central America, including Costa Rica, which is where I saw them. I love Costa Rica; it's one of my favorite places to visit for the food, the wildlife, the people—all of it. Black-faced grosbeaks are gregarious and noisy birds that make the country even more magical. They live in rather large groups and feed on insects, seeds, fruits, and nectar. When it comes to raising chicks, it isn't just the parents that dive right in. Multiple adults chip in by bringing food to the nest and feeding the chicks.

Some parents feel as though they have lost themselves and the individuality they had *before* they were parents; as though people now only relate to them as parents, not as individuals. That can be a challenging adjustment for some mothers and fathers. I think that by incorporating a grosbeak strategy of building a solid network and asking for and receiving help in a multitude of ways (from relatives, nonrelatives, fellow parents, and nonparents), parents can keep a strong bond with each other and raise children more successfully.

WILD LESSONS

* With better transitional support for new parents, fatigue, infrequent showering, wardrobe malfunctions, and a suite of other daily basic-care pitfalls might be avoided!
* Those parents that coordinate parenting duties like long-tailed bushtits will reap the benefits by being less sleep-deprived, less irritable, and consequently their infants or children will get more of what they need.
* Neglecting the relationship you have with your partner and failing to maintain time for intimacy can have dire consequences for your relationship. Don't forget to get some cockatiel adult cuddle time in!
* Everything changes for your friends that don't have kids, too, and sometimes they want to help and get some mommy experience. Let them!
* Create a group like the black-faced grosbeaks and ask members to help with other tangible tasks.

CHAPTER 5

Parenting

Breast Is Best and Other Controversies

Not having kids myself allows me to observe, with detachment, how all of my friends and acquaintances parent their children. It makes for a truly fascinating study in behavior. There is a wide range of attitudes, behaviors, and approaches. Some first-time parents tackle raising their child with a high degree of confidence, while others seem completely uncertain about even the tiniest decisions. Then there are clashes with other parents, friends, and family members over differences in how to parent. These arguments get heated, and their divisiveness can surpass even the most contentious discussion on politics or religion.

As an outsider to the parenting club, I first became aware of these battles during an unforgettable lunch with my friend Leila. Leila is a remarkable person, with a passion for knowledge and education. When we met up for lunch at a local pub across from the university where we both worked, she was heavily pregnant with her second child. Shortly after sitting down, I noticed her face seemed strained more than usual. She was also preoccupied with her phone, which was out of character for her. Concerned, I asked her what was troubling her. Perhaps relieved at the opportunity to vent, she launched into a passionate diatribe directed at her in-laws faster than a chameleon's tongue catapults itself in pursuit of a grasshopper.

There was a power struggle afoot. Both Leila and her husband approached parenting with the goal of supporting and encouraging their daughter's gradual climb toward independence. At almost eighteen months old, Erin, their first child, was independently mobile, tottering around a bit like a Weebles wobble toy but still proudly managing the stairs with limited help. And not unlike others her age, she held her own cup and fed herself to some degree. With both Leila and her husband needing to work shortly after Erin's brother was due to arrive, they decided that having family help was a better option than daycare.

What was the problem? Unfortunately, grandma and grandpa, by babying her, were systematically undoing all of the developmental steps Erin had achieved. They carried her everywhere and prevented her from doing things on her own. Leila was livid. Despite family meetings and calm discussions about how they wanted to handle childcare matters, grandma and grandpa had their own ideas and insisted on doing things their way (or the "right" way, as they put it). The situation had become so contentious that Leila was contemplating issuing an ultimatum: *Raise our daughter the way we want or leave our house.*

Of course, over time I came to understand that letting an eighteen-month-old do what she can on her own is only one of many of the parenting controversies dividing families and even friendships. Around breastfeeding alone there is a laundry list of issues: whether or not to breastfeed, doing it in public or private, and for how long. On the flip side, you have parents that, unlike Leila and her husband, feel hesitant about whether or not they are doing this whole parenting thing properly. The war is not with other parents or family members, but rather with the voice in their own heads as they question their choices and feel guilty for everything. As more of my friends became parents, and I became all too familiar with the "wars" waged on the battlegrounds of parenting, I began to wonder whether my research could provide any insight. Naturally, these are battles that can have no winner, so I thought it best not to put myself in the line of fire, so to speak, and rather than specifically advocate for doing this over that, I decided instead to tackle some of these matters as new avenues into my own studies—to see how animals deal with these same situations.

What's All the Fuss over Milk?

It seems appropriate to start things off with what is arguably the most natural thing for us to do: feed our children. As mammals, we suck milk from our parents, usually mothers (can't forget about those Dayak fruit bat dads). And yet something so simple and straightforward is riddled with landmines. *It's the best. It doesn't matter—do it whenever, wherever. Do it only in private. Do it for three months. Six months. Eight years.* Yikes! Before we even begin to tackle the question of how to feed our young, let's first expand our perspectives. And since parenting is anything that increases the chances that our offspring survive, there is a wide range of options out there when it comes to something as fundamental as feeding newborns.

One somewhat, shall we say, unconventional approach is to eat your mother. The technical term is matrophagy (mother-eating). In a bizarre twist on who's coming to dinner, a desert spider mother goes to the extreme to get her kids off to a great start. These strange-looking spiders can, as their name implies, be found in desert climates. You can imagine that finding a place to build your web could present some challenges, but there are plants in the desert, and those plants are where you will find these spiders.

After being serenaded by the male with delicate vibrations on her web, the soon-to-be-dessert spider mom clutches her sac of fertilized eggs in a cocooned silk ball near her mouth. While the mom is protecting her encased sphere of soon-to-be baby spiders, she continues to eat a nutritious menu of various insects. Then when the spiderlings are ready to hatch, the mom helps them escape their wrapping and begins regurgitating the pre-eaten food. The thing is, this also triggers a waterfall of digestive enzymes, which pour into the mothers' system and slowly liquefy her from the inside. This goes on for about two weeks, and all the while she is protecting and feeding her young, until she dies and they eat the rest of her before striking out on their own.[1]

Admittedly, that is a bit of an extreme form of parenting, but it does have its advantages. In species where matrophagy occurs, the offspring have an enhanced chance of survival because they are able to grow larger before having to leave home. Plus, their mom is still around for about two weeks to protect them—though let's be honest, her ability to do so will diminish daily as she begins to disappear into their bellies.[2] In experiments where the offspring are removed from the mother and not provided with additional nutrition, the survival rate plummets.

Not to be outdone, a limbless snake-like amphibian called the Taita Hills caecilian, found in the Taita Hills of Kenya, feeds her offspring in an altogether different way. She lays her eggs, and when they hatch, the babies are equipped with razor-sharp baby teeth. Don't worry, the hatchlings don't eat their mother. Well, not exactly. While her babies are developing, her skin is preparing itself. The outer layer becomes thick, bubbling with fats and protein much like the milk we produce. When the mini caecilians emerge from their shells, they use their little teeth to tear off pieces of her skin. As you can imagine, babies are typically voracious eaters, and these are no different. They gobble up every last shred and morsel of the nutritious skin. By the time they are ready to leave the nest, the mom has lost about 14 percent of her body weight and is a tad bit paler.[3]

Nursing or feeding offspring expends a good deal of energy no matter how you do it, but the emergence of true lactation was a pivotal moment in mammal evolution, allowing for increased survival of altricial young via the delivery of nutrients past gestation and providing a means of directly transferring immunological competence. And it's not just placental mammals like us that have mammary glands that produce milk for our offspring. The echidna, or spiny anteater (not closely related to regular anteaters), is one such non-placental mammal. These animals used to get a bad rap; I remember my graduate advisor remarking that echidnas are probably one of the only truly random foragers that exist. Most animals have some kind of plan when they go in search of food, even if that plan is to avoid becoming food for someone else. And so I always imagined that the echidna wobbled around awkwardly, unsteady on its little feet as it fumbled through the forest, hoping to bump into food. As it happens, they are hardly staggering about like drunken fools. It has since been discovered that they have special cells in their long noses that can detect the electric signals given off by insects—so they follow their noses, after all, to guide themselves to their meals.

To return to the subject of nursing, echidnas belong to an unusual group called monotremes—though they lay eggs, they are mammals because they nurse their young (called puggles), lactating via mammary glands just like us. Their milk is made up of casein, fat, ash, fucosyllactose and difucosyllactose. Difucosyllactose is also found in human milk and is well known for its ability to protect against infection and disease. It also promotes the growth of important probiotic, or helpful, bacteria.[4] Thus, despite the vast difference between ourselves and echidnas, the food we make for our babies is remarkably similar.

Even though the proportion of fat, protein, and other compounds that make up milk varies from species to species, the benefits are largely the same: nutrition, bonding, and immunocompetence, the latter being perhaps the most critical. Fascinatingly, getting the full protective benefits requires the mixing of a baby's saliva with the milk. I first read about this on a fantastic blog called *Mammals Suck . . . Milk!*[5] That led me to some recent research revealing how both breast milk and baby dribble contain high levels of an enzyme that, when combined, produces hydrogen peroxide in great enough quantities that it halts the growth of staph and salmonella while simultaneously promoting good probiotic gut bacteria! Thus, milk and baby drool deliver quite a punch to potential bacterial infection that formula cannot match.[6] I will never look at a drooling baby the same way again.

The high levels of that same enzyme in baby drool are now thought to help prevent mastitis, or inflammation of the breast, which can be caused by blocked milk ducts. Some experts mistakenly tell mothers that bacteria from your baby's mouth can cause this painful condition. Clearly any doctor that suggests this isn't keeping up with the research. Although there is certainly variation in the levels of protective enzymes in both babies and their mothers, I would suspect there is a relationship between receiving antibiotics during labor and the probability of developing mastitis or even thrush. There are other reasons milk ducts can get clogged, but none place the blame on the baby!

One of those reasons is milk stasis. If milk isn't flowing and stagnates, clogging happens, leading to mastitis. And according to Barbara—the anthropologist with the odd potato craving and who, incidentally, tested positive to group B streptococcus (strep B) and received antibiotics while in labor—unclogging a nipple was arguably more painful than labor. But why would milk not flow? In some situations a mom has difficulty helping her newborn latch on and suckle. Sheila, with the big-headed baby, had another evolutionary question once she was attempting to nurse her son for the first time: "If breastfeeding, which is undertaken by all mammals, is so beneficial to the child, then why do some of us have so much trouble with it?"

Human babies, together with all the great apes, and baby quokkas (a marsupial), and a suite of other species can move themselves toward the nipple unassisted as soon as they are born. Yup, if you let a human baby rest on the mother's chest, the nipple is like a beacon of light flashing the promise of food and other goodies to a newborn, and he or she will reflexively move toward it unassisted. Sheila had the benefit of a lactation consultant. Residing in Switzerland, she was lucky to live in a country that provides mothers with a range of support services. Professionals even come to your home and help out!

Still, there can be difficulties. Sometimes this is the result of positioning. Because of the difference between how human infants suck compared to other species, the nipple has to be positioned just so. Their suckling instinct is strong, but a poorly positioned nipple will frustrate everyone involved. We are not alone in our struggle. We share with our primate cousins—such as chimpanzees, baboons, macaques, and tamarins—the need to learn *how* to breastfeed. How do they learn? By watching others. This is particularly true among all the great apes (chimpanzees, gorillas, orangutans).[7]

The shape of the human breast and the different sucking action of human infants means that we, as a species, have the greatest difficulty (compared to other primates) nursing our infants. As with many things, it comes down to technique, which moms have to learn. The baby gapes and then the nipple has to be thrust deep toward the back of the mouth, which requires mom's help. If this doesn't happen, then milk flow is reduced and you get cracked, bleeding, clogged nipples.[8] Ouch. I don't know when or where we got the message that nursing our babies was this magically, automatic, effortless process. For some, it truly is. But many moms experience difficulty, which can lead to feelings of inadequacy and anxiety because of the misconception that it *should* be easy. This creates unnecessary psychological trauma and sets new moms up to fail at one of the most important tasks: feeding their babies!

It is well accepted by evolutionary biologists like myself that, just like other primates, we need to learn how to nurse our young from others. At the same time, even human mothers who have someone to teach them and then successfully begin nursing have to confront something no other species has to contend with: social opposition to unrestricted breastfeeding. We are the only animal that has "issues" about this. One option: Do as my friend Barbara did. Barbara, a teacher, took her baby into the classroom, and when her daughter fussed she first tried to ignore it, but ended up popping out the boob, sticking it in her daughter's mouth, and continuing to lecture without missing a beat. All of her students applauded her, and not because little Emily stopped fussing.

For those of us whose obstacles to breastfeeding are insurmountable, we are fortunate that we have an alternative option. If a chimpanzee mother cannot successfully nurse her baby, it will die. This was certainly true for humans before the development of formula, unless they were able to have a wet nurse or had access to alloparental care (care from nonparents—more on this in Chapter 9). Around the world, in many other countries where formula is not within easy reach, over 98 percent of mothers breastfeed for six months to two years and beyond. In the United States and other Western countries, rates start out lower and fall precipitously to 55 percent after a mere six weeks. This contradicts all public health and World Health Organization recommendations (and incidentally evolution) to breastfeed for substantially longer periods, with six months an absolute minimum to gain health benefits. What gives? Our generally too-short period of breastfeeding is likely due to a combination of marketing campaigns promoting formula, the lack of training in how to breastfeed, social (and employment) restrictions on feeding as frequently as is needed, and the outright hostility and shaming that many mothers experience.[9] There is a serious cognitive dissonance at play if we insist that breastfeeding is optimal (which it is), demand that women figure it out on their own, and subsequently crucify, restrict, oppose, or otherwise humiliate mothers who do just that.

Let's say that, like Barbara, you manage to ignore all of this and, assuming you are able to, you soldier on with breastfeeding. How long are you supposed to do that? The answer is not straightforward. If we look at other species, we see there is a ton of variation when it comes to how long they breastfeed. What we do know is that the duration depends largely on the developmental patterns of young.

For instance, hooded seal moms don't have to organize their lives around nursing their pups for very long at all. Hooded seals, which belong to a group called earless seals (considered to be true seals), are a bit unusual. They're able to withstand deep dives, divert blood to a layer of blubber under the skin, and streamline bodies for efficient long-distance swimming. Male hooded seals have another feature, one that is the source of their name: a stretchable nasal cavity that can be inflated and deflated. It also makes a ton of noise.

When it comes to nursing, hooded seal moms lactate for a mere four days, the shortest for any mammal. But within those four days their milk is like Wonder Woman milk. The pups gain an average of ten to twelve pounds per day! This lightning-speed growth rate can be attributed to the fat content of the milk—a whopping 60 to 70 percent—and the fact that pups lie around basking in all the fat they're consuming. Unlike with most other seal species, where the moms come and go, hooded seal mothers do nothing else for those four days except feed their pups.

Why do they nurse for such a short period of time? Because they give birth in the icy waters of the North Atlantic and Arctic oceans, pups are born large and with a layer of blubber in order to withstand this extreme environment. That means the mom has already invested an enormous amount of energy to produce a pup that is almost 15 percent as large as her. That would be the equivalent of a 21-pound baby being born to a woman who weighs 140 pounds. Being born with all that fat is a trade-off allowing for the nursing period to be short but intense, where pups consume an average of fifteen to seventeen pounds of milk per day and convert almost all of it into body tissue. This allows mother hooded seals to invest the least amount of time to nursing after birth compared with all the other true seals.[10]

In stark contrast to the seals, we have the orangutan. The first orangutan I met, while volunteering at what is now the Center for Great Apes in Florida, was a three-year-old named Ruby. At the time, the sanctuary was temporarily located at Parrot Jungle (now Jungle

Island). Ruby belonged to the then-owner of Parrot Jungle and was frequently on display playing on the jungle gym under human supervision. She was a pistol and loved being a baby. On more than one occasion, she flung herself over the barrier to snatch a bottle away from an unsuspecting human infant or toddler. Her penchant for bottle-snatching was not terribly surprising when you consider that orangutans nurse their offspring approximately seven years longer than any other land mammal. At three years old, she was hardly inclined or ready to stop drinking from a bottle.

Why do orangutans nurse their kids for so long? Because they have one of the slowest growth rates of all the great apes, with females not having their first baby until they are about fifteen years old. And they have enormous brains. Most animals that have big brains, including humans, take a longer time to mature and require multi-year lactation because the young don't develop fast enough to take advantage of food when it peaks in availability, as is the case with orangutans, whose diet includes a high proportion of fruit. Although they begin transitioning to solid food around one and a half years old and their intake steadily increases with each passing year, they would starve if they were forced to rely exclusively on the unpredictable fruit supply.[11] For orangutans and other species, weaning is a *process*, a gradual transition. By the end, what little nursing that is done is more about comfort and bonding than nutrition.

Given that we have hefty brains and develop slowly, should we be behaving like other similarly big-brained mammals and nurse for multiple years? We can observe that a substantial portion of the world's population nurses their kids for two to three years. Now some people want to go about nursing as long as an orangutan, and to that I can only pose the question: Is there an objective way to figure out what makes sense for our species based on our biology and how we compare with others?

A personal hero of mine, as far as scientists go, is Tim Clutton-Brock. His research and scientific papers formed the backbone of much of my own work on social behavior. His brilliant work on a variety of species and topics has been an inspiration to me, since I am loath to examine only one question or a single species for my entire career, and he is a scientist who has shown me I don't have to. Back in 1985, he and a colleague asked a very interesting question: Can we predict the duration of things such as pregnancy and nursing based purely on body size? This is a tough question to tackle because there are a lot of factors inherent in body size.

For example, we know that the bigger you are, the bigger the brain you have, the longer the pregnancy, the bigger the baby you have, and the slower that baby develops. But Clutton-Brock and his fellow researchers uncovered something very intriguing: The average age of weaning across a range of primates (including humans) increases as moms become larger. It is an allometric relationship, one in which an animal body changes shape in relationship to body size. When it comes to nursing, this relationship is so strong that a formula can predict what the average weaning age should be for over a hundred primates (including us) . . . if you know the mother's weight. Here it is:

$$\log (\text{weaning age}) = \log (2.71) + 0.56 \text{*log}$$
$$(\text{adult female body weight in grams}) \ [12]$$

It should be noted that this equation doesn't produce wildly different weaning ages. For example, if you are an extremely overweight human, you're not adding years to your nursing time. Rather, the predicted age varies by about a year. Does this formula match the data we see across human cultures? Plugging adult female body weights ranging from 110 to 150 pounds (converted to grams) yields a weaning age of 1.95 to 1.97 years old. Across the globe this corresponds nicely with the data that mothers nurse their kids until approximately two years old. The exceptions, or outliers, are Western cultures like the United States, where fewer than four months is the norm. This is to the detriment of our children and has long-term health implications for them. At the same time, there is little empirical support for treating your infant as though it were an orangutan.

WILD LESSONS

* We are mammals. Mammals lactate, and infant mammals nurse from their moms (or dads—can't forget those darn fruit bats).
* Breastfeeding imparts important nutritional and health benefits that cannot be supplied by formula.
* We need to *learn* how to breastfeed. It isn't automatic—not for us and not for many other primates. Insist on being taught by someone who knows what she is doing!
* When it comes to how long you nurse, there is a range, but it's fairly narrow. Our biology suggests that somewhere between two to three years old is reasonable. If your children are asking to nurse in complete, grammatically complex sentences, you've gone too far.

Crunchy Granola Parenting

As humans, something we love to do, are compelled to do, and carry out all the time is label, categorize, and organize everything in our world. Once we have done that, we aren't finished; we have to judge things as good, bad, better, or worse. And so it goes without saying that there are labels for "better" parenting—for example, crunchy granola, or hippie, parenting. As we have already seen, there's more than one way to be a parent. When it comes to feeding, we explored lactation, liquefaction of the mother, and skin feeding—and that's barely skimming the surface. Obviously, for humans, lactation is more appropriate than skin feeding, but that is because of our evolutionary history, not because there is something fundamentally morally or biologically superior about this mode of parental care compared to others. I mean, it would be morally wrong for you to feed your infant your skin, but I digress. Context matters.

One can slide down this slippery slope by observing something in nature and using that observation as the foundation for moral righteousness, such as nursing your child until he or she is six, seven, or eight years old or beyond, based on a belief that it's superior to do so because that's what orangutans do. As always, I will attempt to avoid this trap as I explore a few of the traits labeled as "granola parenting." I can't cover them all, so I have selected two to focus on: sling-wearing and co-sleeping.

When it comes to slings, I first saw one in 1999, before they went mainstream. It captured my curiosity immediately. At the time, I was at Northern Arizona University getting my master's degree in biology, and one of the faculty had just returned from maternity leave and had her baby wrapped in an unfamiliar contraption. I thought it was quite clever because it kept her hands free. The modern-day sling is not much different from the solution that the !Kung, a hunter-gatherer people living in the Kalahari Desert, have been using for ages: They carry their infants in a sling off to the side.

Now that the use of slings has soared in popularity, a lot of claims are being made about the devices. I've heard, for example, that if you wear one, your baby will grow up to be smarter and healthier. Why? Because the baby cries less, the theory goes, so he or she has more free time to learn.[13] Hmmmm . . . I'm sorry to be the one to break the bad news, but there's no science proving any such link. But the practical benefits are real and worth it. That's why parents all over the world have been carrying their babies around with them everywhere they go, one way or another, for thousands of years. Slings are needed. Why? Because most mothers (and fathers) are not idle.

Well, unless we are talking about sea otters. Sea otters are wondrous and delightful to watch as long as we don't mistake their seeming playfulness for friendliness. The mothers are devoted, to the degree that they often attempt to continue to care for a pup that has died, refusing to abandon it. Until they are about a month old, pups don't swim. This makes life challenging for the moms, since sea otters live in the water. Thus during the first month, the moms spend the vast majority of their time resting, nursing, or grooming their pups while floating on their backs so that they can carry their babies on their stomachs.

With little to do, pups under a year old spend the first month mostly sleeping and eating. But the moms have to eat as well. Some mothers have been observed wrapping their young pups in seaweed to keep them afloat while they dive for food, and sea otter pups have a special layer of thick fur that works to trap a layer of air, keeping them afloat during this brief period of time.[14] But sometimes, while left alone, they drown.

It's plain to see we are not the only species to have to find a way to solve the problem of how or if to bring our babies with us everywhere we go. When you think of wolf spiders, motherhood may not be the first thing that jumps out at you. And wolf spiders do jump toward you. It's pretty scary. They can also be big and furry.

At any rate, they are active mostly at night, which makes observing some of their parenting behavior a bit challenging. Nevertheless, observations of the dotted wolf spider reveal that wolf spider mothers are lugging their babies around before they are even born. The egg sac that houses a mother's precious little ones is attached to her abdomen. The mothers are so devoted to their developing offspring that when a researcher, in what I think was a cruel experiment, cut the egg sac and opened it up to take a peek, the mother spider, upon having her scattered eggs returned to her, tried in vain to collect them all and re-spin a sac. When this proved futile, she simply covered them with her body until they hatched.[15]

However, it is once the spiderlings emerge that the real fun begins. One at a time they make their way out of the sac and clamber up onto their mother's "back," which is really the backside of her abdomen. As each spiderling finds a secure place, it flattens itself and hangs on. All fifty to three hundred brothers and sisters ride on mom's back for several weeks. The mother doesn't eat, but she and her babies will drink water together and use her legs as ladders to scamper up and down.

Human mothers in Japan also traditionally carry their babies on their backs, but in an *onbuhimo*. Onbuhimos consist of a small rectangular piece of padded fabric with ties that wrap around and crisscross over the parent's chest. Traditionally, they are lightweight and compact. I found out about these during my undergraduate years, though not, as it happens, in the classroom. I was working at several Japanese restaurants—at that time sushi was an expensive novelty, so Japanese restaurants gave me the biggest return on my time. While I was working at Fuji's, I met Noburo. He is from a small island in the Okinawa prefecture of Japan, and his family grew tulips that they shipped to the Netherlands. He was, and still is, one of the top sushi chefs in South Florida. We became fast friends and roommates, living together for six years.

Shortly after meeting him, though, I noticed something odd about the back of his head. It was flat as a board. Never one to miss an opportunity to ask an inappropriate question, I inquired about his extraordinarily flat head. He informed me that rather than use an onbuhimo, where the infant's head is free to move about, his mother made her own out of a wooden board and cloth straps. His head was against the board while she worked the tulip fields. Until then, I never knew the skull could be shaped in this way, but of course it makes sense. It takes a while for the cranial bones of an infant's skull to come together. The six bones of the skull remain separate but connected by sutures to accommodate the rapid brain growth that is taking place in the first one to two years. Constant pressure on the back of the head can cause these pliable bones to flatten out. This can also happen for other reasons, usually because a baby is kept lying down on his or her back in a crib, in a playpen, or on the floor too much of the time. If you catch it early enough, the flattening can sometimes be corrected, but your baby will have to be outfitted with a special medical helmet that they have to wear constantly.

Of course, some mothers prefer or need to carry their infants on the front, not the back. Despite most newborn infant primates being able to grip tightly onto their mother's or father's hair immediately after birth, withstanding the bouncing, leaping, or running through treetops or on the ground while riding on the back proves too difficult until they are older. That's why most primates carry their newborns on the front of their bodies and can be seen supporting their infant with one hand, at least initially.

The evolutionary remnant of this ability to grip is possessed by human newborns as well. Often people will gasp in surprise at the seemingly superhuman strength infants exhibit when they curl their fingers or toes around something. Whether the baby is a boy or a girl, big or small, use the left hand or right, this reflex, called the palmar grasp, gets substantially stronger over the first seventy-two hours of life.[16] Of course, if we humans hadn't gone and lost the majority of our body hair for other very good reasons, then we'd probably still be using this tight-grip strategy, and our babies could cling to us with minimal support. Unfortunately, that isn't the case, so we've had to engineer other mechanisms and contraptions to do the job for us.

Other animals don't have the ability to make a sling, carrier, or onbuhimo to transport their offspring, but one approach we do share with vervet monkeys is getting someone else to carry your baby for you. Genius! I wonder if this is why, after some reasonable amount of time, mothers are always saying, "Here. Do you want to hold her/ him?" Admittedly, most new mothers of all species are reluctant to let anyone, especially a nonrelative, near their babies, much less carry them, at least initially. Chimpanzee mothers are loath to hand over their infants before they are almost four months old. But after about a month, vervet monkey moms let younger, inexperienced females do some of the heavy lifting.

Vervet monkeys are small Old World monkeys found throughout Africa, and they have made important contributions to our understanding of the evolution of social behavior. My experience with vervet monkeys, however, is outside of the research arena. It's a bit more personal. On a visit to South Africa, I had the opportunity to visit Kruger National Park. It was a dream come true. Upon entering the park, you are given strict instructions: Don't leave your windows down, don't get out of the vehicle (seriously, don't ever do this), be in the campgrounds by dusk, and do not under any circumstances feed the animals. Pretty straightforward. Not complicated. I saw lions, leopards, hyenas, baboons, a rhino, elephants, and, among other animals, vervet monkeys. People seem to disregard those instructions, especially the feeding one, to their own detriment and to that of the animals—not to mention to other people, like myself.

Here I was, with my newly minted master's degree, confident about studying animals in the wild, and I nearly peed my pants when, at a rest stop in the park where it was safe to exit the vehicle, I sat down to eat my sandwich and was confronted by a screaming vervet monkey. Standing on the other side of the table, the monkey snatched my food and glared at me, defiant, challenging me as if to say, "Now what are you going to do about that?" I did the only sensible thing. I let the monkey have my sandwich.

A mother vervet can typically move around freely while carrying her baby, since it grips her tightly around the abdomen while she travels on all fours, or, when it grows a little older, rides on her back. Still, carrying all that extra weight gets tiring. Perhaps that is why about half the time that another female tries to hold or carry a mom's infant for her, the mom allows it.[17] But she won't permit just any female. Older subadult females, equivalent to teenagers, are allowed to hold or carry infants more often than younger ones, probably because they are better at it. The benefits extend to the females that want to carry infants around because they gain experience and practice, presumably making them better moms when they finally have a baby.

In this way, vervets and traditional hunter-gatherer societies are very similar. Babies are constantly being carried by someone. Everyone in the community is on the move, and often a new baby is not being carried by its own mother. And whatever the method, there does come an end to all this carrying. Chimpanzees carry their kids around until they are three or four. That's because young chimpanzees do not just have to learn how to navigate terra firma, they also have to build up the confidence and strength to climb and move, sometimes rapidly, through the trees. It would be the difference between a toddler walking or running extremely fast without falling down. What does this mean for us? We can see that although it is unrealistic for parents to do all the heavy lifting when it comes to carrying kids around (for both humans and other animals), offspring are still held and carried the vast majority of the time. For humans and other primates, this goes well into "toddlerhood." How do parents accomplish this? By enlisting the help of friends and family.

I remember when I went to help a friend out after major abdominal surgery and her seven-year-old son had fallen asleep on the couch and she wanted to put him to bed. She looked at me expectantly and I said, "What? You want me to wake him up?" She laughed, saying, "No, I want you to carry him to bed." "Oh, I see," I replied. And so I hoisted up this little—well, not so little—person, and as I stumbled toward his bedroom with him awkwardly draped across my arms, I kept thinking, *He looks so peaceful sleeping. How sweet.* But I was also thinking, *This kid is totally capable of walking and I want to wake him up right now because he is heavy!* Thus, there is an age limit. Once kids can walk properly, the amount of time they are carried begins to decline. And there is a difference between holding and carrying. So cuddle and hold all you want, but stop carrying your six-year-old around!

Holding and cuddling brings us to sleeping, or co-sleeping. Most animals that provide substantial post-birth care sleep with their kids. Bird parents don't have a separate, adults-only nest, and grey mouse lemurs don't leave their young babies to fend for themselves in a different tree hole. That being said, babies need a lot of sleep, and sometimes the mom isn't around while they are taking a nap. It happens all the time with deer fawns; when people stumble on a little fawn, many erroneously believe it has been abandoned and "rescue" it. The distress and panic the mother experiences when she returns to the hiding spot only to find her fawn gone is not something we can quantify, but I am certain it is not trivial.

Sometimes, as in the case of the sea otter, it is a matter of necessity to sleep with her baby since, as we've learned, baby sea otters can't swim and therefore sleep on their mothers for quite some time. We don't know whether, if given a choice, mother sea otters would place their pups on something else to sleep through the night. What we do know for sure is that, for many other animals, sleeping away from one's mother raises cortisol levels in *both* mother and offspring. I won't go through all the awful experiments where infant primates or baby rats were taken from their mothers and prevented from sleeping with them, but they revealed how important co-sleeping is for parents and offspring. Does that hold up for us? There's no reason to doubt it.

Now how *long* the young sleep with their parents is another matter altogether and has a lot to do with how long it takes for them to grow up. A stellar co-parenting species is the humble California mouse. These mice make their homes primarily in chaparral and scrub regions of California from San Francisco all the way down to the Baja peninsula. Males and females form a pair bond for life, with life being under two years. During their brief stint on earth, they live in somewhat permanent family groups, where older kids still hang around in the nest with their mom, dad, and the new kids. The moms and dads regularly huddle over, sleeping with the youngest pups, but do so less over time. By the time baby mice reach thirty days old, both moms and dads spend only about seven hours in the nest with them. And what about the night, when California mice are more active? What happens then? For the first several weeks, the mom and dad play tag team so that at least one parent is in the nest with their young.[18]

The California mouse sleeping arrangement mirrors how nearly all humans before the turn of the nineteenth century handled sleeping arrangements. To this day, if we exclude most westernized cultures, co-sleeping with children is the norm around the world. Indeed, it's rooted in our evolutionary history for parents to sleep with their children. One question we need to always be asking ourselves is whether or not practicing new parenting behaviors that run counter to our biology and evolutionary history harm our offspring.

Forgoing evolutionary considerations in something so simple as sleep is creating a mismatch between what infants can adapt to and deal with and what we have decided culturally is the "best" practice. And it may just be why sleep troubles in infants and young children is the biggest parenting problem reported in the United States and other westernized cultures but is virtually absent anywhere else *in the world*.[19] And the problem continues into adulthood, with about 60 percent of Americans reporting difficulty sleeping.

Though I rarely pull out the "look at what our closest primate cousins do" card, in this case it is appropriate. We share many features with chimpanzees, including our DNA, transmission of cultural practices, social behavior, and development of our young. As already mentioned above, like other primates, chimpanzee moms carry their infants, first in front, then on the back for three to four years. Several films made about Dr. Jane Goodall's work on the chimpanzees of Gombe devote time to highlighting the sleeping situation of chimpanzees. Baby chimpanzees sleep with their mothers exclusively until they are around five or six. In the meantime they practice their nest-building skills, so mothers generally do not have to kick their kids out of the nest; it just naturally happens.

Some people erroneously believe that if they fail to get their baby sleeping alone in a room by him- or herself right off the bat, they will end up with a needy, dependent child who will refuse to sleep alone forever. We can credit Dr. Ferber for promoting this misconception, and once again we can debunk him, because the opposite is more accurate. There is solid evidence that infants who co-sleep with their parents from the start feel more secure, confident, are more self-sufficient, and are more successful in making friends.[20]

There are a host of other benefits as well. First, children who feel secure will be motivated to become more self-reliant. Second, breastfeeding and potty training are easier to accomplish for parents who are sharing the same bed or room with their infants (or children) and this results in more sleep for everyone. And the research shows that the rate and duration of breastfeeding goes up with co-sleeping.[21] But perhaps the most astonishing result of co-sleeping is that it can reduce the incidence of SIDS (sudden infant death syndrome). SIDS is virtually nonexistent in other parts of the world where bed sharing or co-sleeping is the norm.[22-23] Part of this may have to do with the feedback happening between mother and infant and her unconscious alertness and attention to her infant, even if the infant is simply in the same room. Also, babies that sleep with their mothers are more likely to sleep on their backs rather than on their stomachs because that position naturally offers easy access to the breast during the night.[24] Infants that sleep with a parent also have higher body temperatures and are neurologically primed to respond to a sleeping partner.[25] We evolved to sleep with our mothers; all of our biological and neurological systems demonstrate this.

However, it is important to realize that, despite the evolutionary underpinning, this does not imply that all forms of bed sharing are equally safe. If you smoke, drink, do drugs, or are sleeping on inappropriate surfaces, the risk of harm to your infant can go up. These other factors may ultimately explain how some experts reach the conclusion that co-sleeping is dangerous.

WILD LESSONS

* Using a sling or a back carrier to hold your baby and still get other stuff done doesn't make you a hippie parent. It makes you a parent. And it won't make your baby smarter or dumber. Quieter, perhaps.

- If you really want to free yourself up, have someone you trust carry your baby around with you all day long (just make sure it's not a vervet monkey).
- Once kids can walk properly on their own, let them, but continue to provide many opportunities for them to be held and cuddled by yourself and other trusted adults.
- Human infants are evolutionarily adapted to sleep with their parents and come complete with sensory systems attuned to sleeping *with* someone.
- It is the norm across much of the world to share our sleeping space with our young.
- The cultural ideology of "self-sufficiency" has given rise to the myth that feeding and sleeping must be rigorously scheduled and sleeping must be done alone.
- Not all forms of co-sleeping are safe! Do your homework and learn how to safely sleep with your infant to maximize his or her physical and physiological well-being.

Staying at Home Versus Being a Working Parent

This chapter is titled "Parenting" because even though as a society we throw around the phrase "working mom," we really should be talking about being a "working *parent*." If we are going to investigate whether moms should be working, then, given that we have evolved to be a largely biparental species, we need to apply the same logic to whether or not men should be working. As we have already seen above, relatively recent cultural norms do not necessarily reflect biology. Therefore, as I frequently do, I suggest we set aside cultural conventions and beliefs and explore, in a gender-neutral fashion, the question of staying at home versus working—because as far as the offspring are concerned, conventions are irrelevant.

Working parents are not a new phenomenon. For much of human history, mothers and fathers have worked. The nature of that work and who does what varies from place to place, from culture to culture, and depends on the time period. For instance, for the !Kung, mentioned above, technological resources are extremely limited, and they rely on each other and the contribution of each community member to survive. Women traditionally are the gatherers, but this is not a matter of strolling through lush forests collecting berries that are abundant and practically falling into their baskets.

No, in the harsh environment of the Kalahari Desert, the women may walk many miles from camp, carrying their infants to go collect nuts, fruits, and tubers to bring back. The men hunt, make tools, and perform general maintenance. To those more at home with post-1950 American culture, it may sound more natural for men to go off to work and bring home the proverbial bacon, but the reality is the meat is hard to come by and the entire !Kung community relies heavily on the food that *women* bring back. And !Kung dads spend a tremendous amount of time playing with and watching over their children compared to the measly one hour of contact per day that American men now spend, on average, with their infants.[26]

When it comes to shared, hands-on parenting in the animal kingdom, lions are fascinating because of the extended family that is available to cubs. In most lion populations, the pride is made up of related females and all their kids. The males are transient and can be part of a particular pride anywhere from two to four years, forming coalitions (meaning several males share a group of females). Male lions have gotten a pretty bad rap as dads because of all of the cub killing they do. And it has been a bit overemphasized that females do all the hunting and males just sit around looking gorgeous. It's true that they're beautiful, and when you see a male lion in the wild, resplendent with his full mane, you can't help but feel dazzled even as you sense he could and would rip you to shreds. This isn't to say that females couldn't do the same, but you get a greater sense of the lion's fierceness, power, and strength when you see a male trotting effortlessly by or even just yawning. At any rate, males do hunt, and how much they hunt may depend on the location and what they are hunting.[27] The other misconception is that males sometimes kill cubs but never care for them. But fathers don't kill their own cubs; they lounge around with them and are remarkably tolerant and even downright playful with them sometimes.

You might be inclined to say that how societies like the !Kung and animals like lions live has nothing to do with us. But the reality is that, in terms of contemporary society, the "stay-at-home mom" is a relatively new phenomenon. Prior to 1950, most families had working parents, so relatives helped take care of the kids. The American ideal that was marketed so skillfully to everyone was that the goal, the pinnacle of success, was if your wife could stay home with the children. And with this came the notion that this was the "best" way to raise kids. In reality, I propose that it has given generations of Americans a distorted view of what family is and how children ought to be raised.

The reality was (and still is) that reliance on a single wage earner was only for the wealthy and aristocratic. Even for the upper class, this never meant the mother stayed home and took care of the kids. That was what nannies were for. Even today, we still see the same thing. For some, having a nanny means you've achieved a certain level of status, and in wealthy families, even if the mother doesn't work, a nanny or other caregiver will handle the bulk of the day-to-day childcare duties. But I want to focus on parents who work *and* take care of their kids (the reality for most parents).

For this, birds are a fantastic example. Let's zoom in on mute swans for a closer look at one strategy. Even though many of us are familiar with these large, graceful swans—with their white bodies and orange beaks decorated delicately with black, making for a look that reminds me of Zorro—they are not native to North America. I first learned it was a mistake to tango with a mute swan (or any swan really) when I was a volunteer at the Palm Beach Zoo in Palm Beach Florida. I was still trying to figure out what I wanted to do with my life, and I thought maybe working at a zoo was a good option. Much of my time was spent feeding animals and cleaning up their poop. Lots and lots of poop.

One area that I was in charge of contained the tapirs, fascinating creatures that resemble strange pigs. Inside their exhibit was a pair of swans. Every time I went into the exhibit, a large grassy area with the tapirs located toward the back, I had to pass the swans. The very first time I went in I was unprepared for what lay ahead. As I exited the golf cart, I caught a glimpse of something out of the corner of my eye. I turned and immediately saw that a very large mute swan, head low like a sumo wrestler charging its opponent, was coming right for me. As it closed in on me I was paralyzed, more out of surprise than fear, and had no inkling what was about to happen: I got a smackdown from a swan. While I escaped relatively unscathed, being slapped by a swan was a humbling experience. Needless to say, after this first encounter I always came prepared—with bucket in hand to fend off the angry swan until I passed what she or he considered to be a private oasis of grassy knoll.

This fierceness is one of the attributes that makes mute swans great partners to each other and protectors of their offspring. Both males and females take their parenting duties very seriously, as many a canoeist, pedestrian, and this one-time wannabe zookeeper undoubtedly have discovered. Like all parents, mute swans have to get "work" done in addition to their parenting duties. They have to eat, they have to move around to find food (to eat), and of course they have to take care of the nest and protect the territory. The question is, do they divide up all this work along gender roles?

Sort of. Females almost exclusively incubate the eggs, which accounts for a whopping 90 percent of her time, leaving her little opportunity to do anything else except occasionally eat. Her partner is hardly inactive. He's the one running around defending the nest from intruders. Once the chicks arrive on the scene, however, things balance out a bit more. While they raise their chicks, both parents get equal time for resting and cleaning themselves up. Another difference?

As the chicks begin their journey out of the shell, protection isn't just up to the dad; both parents share this job equally too, but males are more likely to get into an altercation. I'm going to guess it was the dad that came after me. Despite sharing many of their parenting duties, somehow it's the mom who feeds the chicks 75 percent of the time.[28]

Unlike in the case of swans, where moms handle the bulk of the feeding duties and dads take charge of protecting the family, in other species, parenting roles are determined by the dynamics between individuals, not gender. This is very true for some cichlid fish and human families. There are at least 1,600 species of cichlid fish occupying a variety of habitats, with the greatest number of species living primarily in freshwater rivers and lakes. For one little species, the Golden Julie, there is no hard-and-fast rule about who cares for the babies, or in this case, the eggs. Both parents take an active role in brood care, which involves fanning the eggs, cleaning the eggs, and, after the eggs hatch, cleaning the larvae. That is because the babies absorb the remaining yolk sack before they are fully free-swimming baby fish, or fry. And the mom and dad still watch over the kids until they are large enough, which takes about three-and-a-half months.

Even though who does what is not set in stone, it is not divided equally down the middle. It turns out that domestic life for the Golden Julie is determined by size. One partner, the larger one, dominates and makes the other perform most of the parenting duties. How do we know this? Because the larger of the pair, male or female, pecks at or headbutts his or her partner into staying at the nest and taking care of the eggs. However, this strategy only works if the size difference is large; if both parents are of similar size, things are more balanced.[29] Fascinatingly, studies have shown a similar dynamic at play in human couples—minus the abusive coercion, thankfully.[30]

As I mentioned, for post-1950s humans, there has been a high value placed on the stay-at-home-mom and go-to-work-dad model of family. This made me wonder whether or not we ever see this scenario in other species. For red-knobbed hornbills this is the norm. I saw my first hornbill on that very same trip to South Africa where I was robbed of my lunch by that vervet monkey. At the time, I knew nothing about hornbills, only that they seemed burdened by their cumbersome, ornate bill adorned with a casque, causing them to appear as if they might topple over at any moment. I thought they seemed self-conscious: They looked at you as though they expected you to crack a joke or make a remark about their gigantic bill. But this is a bird whose neck vertebrae had to fuse together just to lug this thing around. So it must be useful, right? There is still some debate about why they evolved such a structured and layered bill, but some ideas include that the casque, a hollow structure that runs along the upper mandible, serves to reinforce the bill and allow them to exert more force on the foods they eat, which, being omnivores, includes a variety of fruits, insects, and small animals. The beak and casque's being pronounced and colorful probably also plays a role in attracting a great partner.

Finding a mate is an incredibly important choice for hornbills because of this whole stay-at-home mom gig they abide by. For this species, the entire nesting period lasts about 140 days, and for much of that time the mom is literally trapped at home with the eggs and, later, the chicks. They're sealed in the nest for about 108 of those 140 days. Talk about cabin fever! Or, in this case, tree hole fever. Don't worry, the males aren't locking up their females and eggs in a tree hole; it's the females that use their own feces to accomplish this. A female doesn't completely seal herself in; she leaves a narrow slit that acts as a lifeline between her and the outside world. Her mate, the chicks' father, faithfully feeds her, and later the chicks, a diet largely consisting of figs.[31] After the chicks are old enough, the mom breaks out of the nest, visiting infrequently to feed them, leaving the bulk of the parenting duties up to their father until they fledge a few weeks later. Why have hornbills evolved this very unusual parenting approach? Because there is severe competition for nesting holes, so shutting herself in is a way for the female to prevent other hornbills from stealing her young's home.

The script flips the other way when we look at pygmy marmosets, where dads are on the scene before the twins are born. He helps with the delivery, cleans the babies off, and handles all the parenting duties except for feeding, which mom handles. Otherwise, he carries his kids, grooms them, plays with them, and is ever present for the first few weeks. Then, much like us, after the newborns are about two weeks old, sometimes they are "parked" in a safe location nearby while the rest of the group eats! The reason these little primates, and others related to them, have such intense help from males and other group members, is that giving birth is an enormously energy-consuming undertaking. And having twins is even more taxing. Mothers need help, and dads step up.

Clearly, the debate about how much effort each parent should devote to taking care of the kids isn't restricted to human parents. This question, perhaps mirroring our own struggles, has long plagued scientists studying parental care theory. Should the female be the primary caregiver? Should it be the male? Should it be both? If it's both, how should parenting efforts be divided? Animals, on the other hand, don't ask these questions; they just do what they do. So we have an opportunity to observe and try to figure out just why they do things a particular way, given their lifestyles and biology. And as we observe other species, a common theme emerges.

In animals that need both parents, if one parent puts in less effort, the other parent puts in more to partially, but not completely, compensate. There is a benefit to the one who can lower his or her investment and a cost to the parent who must pick up the slack. In human terms, we may think of it like this: If you invest less in your kids, and your partner makes up most of the difference, you may enjoy a more productive career and reap the financial benefits that often come with that. We know how this can play out in humans, but is there any truth to this in animals? Does one partner pay the price for a lower investment by the other? In the case of orange-tufted sunbirds, yes.

Strangely, they aren't orange at all. The male is a metallic green with a purple forehead and copper around the sides of his head. They raise their chicks in intricately woven baskets that hang delicately from branches. In one study, researchers put weights on the tails of some females to see what would happen if they couldn't keep up with their parenting responsibilities. Why would a weight matter? Most passerine birds, like the sunbird, are living life on the edge . . . of starvation. The added weight increases their energy expenditure, and so one might expect that an individual will alter its behavior to compensate. And that's exactly what these females did. They reduced the number of trips they made to feed their kids. Fathers typically take up the role as nest defender, but with the moms not feeding the kids as much as needed, they assisted to compensate. That means they had to work harder. When the researchers did the reverse and removed the dads, the moms compensated by increasing their feeding rate, but they couldn't do it all and were not able to pick up the slack in defending or spending more time at the nest.[32] Whichever way you slice it, both the moms and the dads paid a price when their partners weren't able to assist with the parenting duties.

There seems to be a constant discussion about whether or not mothers should stay at home or work, with the underlying assumption being that children will do better if they are raised in a single-wage-earner household, where the male goes off to work and the female stays home, thus framing a sociocultural question as concern for children. I wonder if this concern is misplaced. Working parents, as I have already mentioned, is not a new concept to humans. What has changed is that we work outside the home, and this constrains our ability to both work *and* be with our children at the same time. Take Your Child to Work Day happens about once a year for most people. When we look across the animal kingdom, lots of species have to figure out how to "work" and raise the kids. Here we have focused on situations where both parents are available. Each species has its own approach to how to divide up parenting when they are in a partnership.

There don't seem to be any hard and fast rules, but one thing is clear: When one parent is slack in his or her contribution, the other parent and the kids pay the price. It is up to the parents to negotiate and work out what will work best for their family. Ultimately, who goes to work, who stays home, and whether one or both parents work doesn't matter as much as how secure and loved children feel and the quality of care they receive. By force-feeding certain cultural ideals, we ignore the real costs to parents—often mothers. One Finnish study revealed that the next most common emotions mothers experience after love are rage and anger. Why? Because they feel isolated, sealed in their homes with their children.[33] Much like a hornbill. Not everyone is cut out for that. If stay-at-home mothering were always the best strategy, it would be common among more species. And it certainly isn't reflective of our evolutionary history.

WILD LESSONS

* All parents are working parents, whether they are stay-at-home hornbill moms or kid-toting tamarin fathers.
* Each couple needs to navigate who does what, but try not to be a domineering or manipulative cichlid about it.
* Staying at home with the kids is not for every parent and can lead to negative emotions and aggression toward one's children.

What's with All the Guilt?

It seems as though everybody and their mother wants to tell moms the right way to take care of their children. With all the judgment being handed out, it can take a toll, leaving parents, usually moms, feeling guilty about everything they are and aren't doing. But perhaps the biggest enemy to mothers is themselves: the constant niggling at the brain like an itch that refuses to be satisfied when scratched. The feeling that somehow, despite all the effort, your best isn't enough. Or maybe it isn't even your best. We have this notion that all care and investment must come exclusively from the mother. But this has rarely been the case in human evolution, and mothers have always moderated how much they give based on the particular circumstances they find themselves in. This is true for all kinds of animals, not just humans.

Even the sleek, beaked puffin moderates how much it can give to its offspring based on how it is doing at any given point in time. Puffins are enchanting, stocky little seabirds that are typically found off of the rocky coasts of the Pacific and Atlantic Oceans. They are expert divers and fly underwater. When it comes to providing for their chicks, how much they provide is related to their own physical condition. When parents are in bad shape, they give less, mostly because they have less to give. How much to invest in a child is a dynamic process that is based, in part, on the condition of the parent. They are incredibly flexible in how much effort they put into feeding their young.[34] If you think about this, it makes perfect sense. We always hear, if you don't take care of yourself first, you can't be there for someone else. Well, the same is true for parenting, though, like puffins, your cup may get pretty darn low before you reduce what you give to your kids.

The cultural myth and expectations we have placed on mothers gives rise to guilt. In many ways, guilt is a moral emotion. Unless you are a psychopath, guilt emerges when you feel that you've done something wrong, usually to another person. The emotion, because it is unpleasant, helps regulate our behavior. It's useful. It prevents us from neglecting our children or being otherwise rotten parents. As we all know though, one can ride the guilt train a bit too far off the tracks and worry that nothing we do is good enough. Or worse, we might feel guilty that we don't feel more guilt about not wanting to give more!

It's difficult to compare the emotion of guilt between humans and animals because we cannot assess if they *feel* guilt. We do not know if mother gorillas agonize over their daily decisions. We know they, along with many other species, experience deep grief and suffering when their offspring dies. But we cannot know whether or not a mother blames herself or feels guilty when something tragic happens.

But in humans, tragedy doesn't have to strike to elicit guilt. And frankly, it's not the big decisions that are the most challenging. It's the hundreds of micro-decisions made every single day. Oftentimes those decisions are made with incomplete information. Life is full of gray areas. As a parent you may choose a daycare, doing as much due diligence as possible, but never feel certain it's the best daycare or that your kids should even be in daycare. You may wonder if you are spending enough time with your children, or, as we discussed above, feel guilty because you work.

But surprisingly, some of the top reasons mothers report feeling guilt have to do with not wanting to give at any particular moment. Maybe it's attention they don't feel like giving at the moment, or time. This typically manifests itself as anger or aggression toward their children (for causing guilt!), a desire to throw in the towel and leave, or a feeling that they favor one child more than another. And then there's this critical source of guilt: not living up to some idealized expectation of motherhood.[35]

What is a mother, or father, to do about all this guilt? Of course the easy answer would be to stop feeling it and experience it only when you are actually being a poor parent. But, again, parenting requires any number of daily decisions, and it can often seem that there's no clear right or wrong thing to do; you just have to decide.

So we can look at how animals make decisions when they have incomplete knowledge. Before any animal can make a decision about what to do, it must first assess the situation. Animals are constantly making judgment calls regarding risk, determining the chance that they might get eaten by a predator while pursuing a new food source. And it is widely accepted that they're constantly making these decisions without perfect information. What happens if we follow their example and let our behavior be guided by a rule-of-thumb approach? It seems sensible and we do this all the time.

The problem is that if we have a standard set of habitual responses to stimuli, we could be way off in our perception of what is the best decision to make in a specific situation. Ants in general are pretty smart creatures and tend to make good decisions. And they have a lot of decisions to make: where to go, which grain of sand to take to build the nest, and where to set up a new nest, for example. House-hunting is an especially big deal for one small species of ant, *Temnothorax albipennis*, found throughout Europe. Ultimately, the ants in a colony arrive at a decision collectively when enough other ants have settled on one particular site after collecting information on several features of each potential nest site. Sometimes, though, when times are difficult and they are stressed, they are under a time crunch to make this important decision and become much less discriminating. And then they end up in a mediocre nest site. When times are good and they don't need to make a decision quickly, they make better decisions.[36] Because this is such a common problem, it goes by the term "speed-accuracy trade-off paradigm" in behavioral ecology.

Animals almost always have inadequate information to make decisions and balance a set strategy with gathering new information and updating their decisions as they go. Stress compromises the ability to make good decisions and to take the time to consider the options. However, the "rightness" of a decision only matters when the magnitude of the consequences if you are wrong are high.[37] In other words, don't sweat the small stuff. And for more consequential parenting decisions, don't trust your gut, take your time.

WILD LESSONS

* Parenting is a dynamic behavior, and if you don't take care of your own needs, the quality of your parenting will suffer.

* Guilt is only useful insofar as it motivates you to be a positive and attentive parent who interacts appropriately with your children and others around you. Beyond that, it doesn't serve a real purpose.

* There is no one-size-fits-all strategy to parenting or making decisions. A rule-of-thumb approach may seem easier, but it increases the chance of making a poor decision.

* Like ants, parents need to make decisions and often lack full information. In the absence of an emergency, it pays to contemplate and assess the options before making a choice.

* Remember: Don't sweat the small stuff. If the consequences of a poor decision are trivial, it won't matter much in the long run, and it's not worth the stress.

CHAPTER 6

We Are Family

How many children to have is a consequential question for humans and other animals alike, and it's a conundrum that parents face whether they have one child at a time or multiple births. Either way, there is a delicate balance. As more offspring arrive on the scene, resources are divided into smaller and smaller shares, often leaving the second, third, fourth, or subsequent child at a significant disadvantage to the others who came before. The conflict between parents and their children, when it comes to time, resources, and personal success, often goes unrecognized. But the trade-off is very real, and having more than one child increases the cost to parents exponentially. Is there an optimal family size? Do we have a responsibility to limit the size of our families? How do other animals figure out how many kids to have in light of how many they can support?

Parents frequently assert that they love all their children equally, but many of us, having all been children at one point, know unequivocally that this does not reflect our own experience. What gives? Are we unaware of how parents play favorites with their children? Are humans hardwired to prefer one child over another? Although the psychological literature is rich with why stress and marital problems lead to favoritism, there is a notable absence of a biological explanation for why we may be predisposed to allocate more time, energy, and other resources to one child over another. What can we learn from other species? How can parents deal with their tendency to play favorites?

Though favoritism by parents can certainly lead to sibling rivalry, inherent competition for access to limited resources sets the stage for a struggle between siblings even without parental provocation. In other animals, this battle is sometimes to the death. We may like to think we are better than this, but violence between siblings is the most common form of domestic violence, with reports that over 60 percent of siblings have engaged in some act of violence! [1] This statistic is not trivial and demands exploration. One way to start is by examining the relationships among siblings in other species in this context. In doing so, we can learn when conflict is most likely to occur, what strategies animals use to minimize it, and how parents can foster more peaceful relations among their children.

At the same time, having siblings is not all bad. Sometimes having a brother or sister is helpful. Lions, cheetahs, and many other species pair up with their siblings to enjoy greater survival and success. Among different-aged siblings, a brother or sister can babysit, help feed the younger sibling, and offer protection out there in the big, bad, wild world. How much do other species rely on the assistance of older children to help raise younger offspring? Let's dive right in and explore the benefits and pitfalls of being a family with multiple children.

Let's Have More!

When people think of armadillos, they generally envision a strange, armored creature looking like part-anteater, part-dinosaur that curls up into a ball when threatened. Although armored from tip to tail, the nine-banded armadillo doesn't roll up like its three-banded counterpart. Instead, it jumps up in the air.

Another unusual feature is that this species doesn't have a lot of choice when it comes to family size. Imagine if every time you got pregnant you had identical quadruplets. The technical term for this is polyembryony. All it means is that one egg splits into two or more offspring. In humans, the probability of naturally occurring identical triplets is roughly 0.00002 out of 1,000,000 births. As for the probability of identical quadruplets? Play the lottery; your odds are better. The difference between us and the nine-banded armadillo? For them, this happens all the time.[2]

This creates a bit of a challenge for the armadillo, and scientists continue to study the phenomenon. How do we explain this automatic set of quadruplets in terms of evolutionary biology, since it forgoes one of the primary benefits of sexual reproduction: the shuffling of the proverbial genetic deck. Instead of four offspring that are some mixture of the mom and dad, each different from the other, for the nine-banded armadillo they are the same mix repeated four times. Genetically identical offspring don't have the benefit of genetic diversity. If something is wrong with one, it will be wrong with all of them. A set of human monozygotic (identical) Swedish triplets studied in the mid-nineties illustrates this point. By the age of twenty, all had developed schizophrenia. The same chromosomal glitch on chromosome 15 was detected, though it was unclear how this defect might have been linked to schizophrenia (research continues in this area).[3]

More recently, a link between an immune system gene, C4, and schizophrenia has been discovered,[4] potentially explaining how genetically identical siblings, or clones, can end up with the same disorder. This example illustrates the dangers of clonal reproduction, which may explain why it is relatively rare in nature. Unfortunately, due to an odd-shaped uterus, the nine-banded armadillo cannot escape this potential pitfall.

Fortunately, many other species, including humans, can and do adjust how many children they have, whether or not they can decide to have one at a time or many at once. Along with all the great apes and many other species—orcas, dolphins, and elephants, to name a few—humans invest *years* in their offspring. This makes it very costly for parents to have children, and each decision to reproduce is a trade-off between current offspring, future offspring, and the parent's own survival. So how exactly can one decide how many children to have?

If you are a Eurasian lynx, you follow the one-size-fits-all strategy. The Eurasian lynx, whose habitat ranges from Siberia to the Himalayas, is the largest of the lynx species. Males roam a territory roughly 150 square miles, larger than the size of Boston, while females require less space, a territory "only" slightly larger than San Francisco—something to think about, the next time you see a lynx at the zoo.

When it comes to making babies, rather than having as many as possible at any one time, the lynx settles on an optimal number that ensures the greatest probability of survival. What is that number? Two.[5] This is independent of how much food is available, how heavy the mom is, the weather, the year, being in the wild, or being in a zoo. Having two cubs instead of, say, four, leads to the greatest survival of a mother lynx's litter under a variety of conditions. This also means that Eurasian lynx are only replacing themselves.

The reason lynx have evolved this strategy may have to do with how they live in difficult environments with a lot of variability in their food supply. Once females give birth, their opportunity to find food shrinks because they have to stay close to the den for the first two months. Because they don't fatten up like some species, they invest less in the size of each cub and how much milk they have to give. It turns out that two cubs is the average limit they can successfully wean at one time. It's a trade-off between quantity and quality.

Ironically, this number of children, two, is also the perceived "perfect" family size for many people. And not just any two: a boy and a girl. A colleague, Margaret, herself from a large family of six siblings, remarked that after the birth of her son, she knew all she had to do was have one more baby, a girl, and the family would be "complete." When I pressed her about what would happen if she had another boy, she shushed me, exclaiming, "Don't jinx me!" As if I had the power to control the gender of her yet-to-be conceived baby. People (or maybe just Margaret) can be strange.

But is there something adaptive or ideal about her position? She, and others, may be onto something. In modern human populations, there is a pretty big trade-off between the number of children you have and the survival and success of those children. Mortality rates go up with each successive sibling in European, American, and contemporary African populations.[6] Many parents decide they want a second child so that their first child has a companion and friend, and can share the responsibility of taking care of them (the parents) when they are older. Having more than one child introduces a lot of variables, not all of them positive.

On the other hand, many people believe that having only one child is a problem, fearing an only child will be a spoiled narcissist. Where did such a notion originate? One place might have been with psychologist G. Stanley Hall, who asserted that singletons, as only children are sometimes called, were afflicted. "The only child," he states, "[is] jealous, selfish, egotistical, dependent, aggressive, domineering or quarrelsome."[7] Indeed, much of the propaganda of the 1920s and '30s reinforced these ideas. This led to teachers, parents, and society at large believing that singletons would be utter failures at home, at school, and ultimately at life.

What does the science say? As early as 1928, studies were reporting modest, if any, differences between only children and those coming from larger families. Where significant differences do exist, ironically, single children are happier, *more* well-adjusted, have higher IQs, perform better overall academically, are more successful, and do not suffer from loneliness, social awkwardness, or unreasonable pressure to excel. Why? It may simply come down to claiming a monopoly on parental resources, being included in more adult activities and conversations, and most importantly, getting all of that undivided parental attention and affection.[8] Despite this, the belief persists that having multiple children is needed to prevent the disaster that will ensue by having just one.

Incidentally, on average eldest children also have a higher IQ than their siblings. These observations have been consistent in human populations, but many have mistakenly attributed these differences in IQ to biological differences as a result of birth order, when it is more likely the result of family dynamics, specifically greater allocation of resources (time, attention, etc.) to firstborns. As if that weren't enough, successive children are also shorter and tend to weigh less at birth.[9]

This pattern, where eldest siblings are larger, is found in bank voles, too. As a general rule, voles are pretty endearing, and the bank vole, the smallest of the voles, with its taste for hazelnuts, is no exception. Like many (but certainly not all) rodents, females can have multiple litters in a year, with anywhere from four to ten pups per litter. Experiments manipulating the number of pups have revealed that the larger the litter, the lower the body mass of pups at weaning.[10] However, the long-term consequences of starting off life a wee bit smaller are less clear-cut and depend heavily on the immediate environment the voles are raised in and the resources available.[11]

This was also the situation in my family—well, except I wasn't part of a litter. As predicted, my older brother's IQ is approximately three points higher than mine, and compared to his healthy 8 pounds, I was a mere 5.5 pounds at birth. Unlike the lynx, and despite the well-documented benefits of having just one child, many parents are having more than one or two children. Is that a problem? And should we humans stop after one child?

For some time now, the human population growth has mirrored exponential growth. What does this mean? It means that we have been reproducing like bigfin reef squid. A member of the highly intelligent group of animals called cephalopods, meaning "head-feet," bigfin reef squid are built for speed. They are superb hunters, thrusting themselves rapidly through the water using jet propulsion and lunging forward with tentacles outstretched in a cone shape to seize their prey.[12]

For a species like this, parents don't invest much in each individual offspring. Releasing sperm and eggs is about it. Heck, the parents may not even know each other because fertilization is largely random via spawning. For many aquatic species, this is the way to go. Males release millions of sperm while females release millions of eggs and hope they cross paths. For them it is about quantity, not quality. Parental investment comes in the form of millions of potential offspring. In ecology we refer to these types of species as "r-strategists." The term itself isn't nearly as important as the reproductive properties of these species. Typically, r-strategists like the bigfin reef squid live in unpredictable environments, have a high mortality rate, develop and reach maturity rapidly, and exhibit minimal parental care.

Despite the herculean efforts to produce millions of baby squid, the world is not covered in bigfin reef squid. Why is that? That's because baby squid are weak and tasty treats for other animals, so the vast majority don't make it to adulthood. And of those that do, life is truly short. That doesn't sound like us at all, yet we belong to the ranks of salmon, bacteria, insects, oysters, and squid in our global rate of reproduction.

Still, there is an abundance of variation in how many children people have. Some people, like myself, have zero children and others may have ten or more. Perhaps, then, there is something else going on. Aside from a common best number, when contemplating how many children to have another strategy is to determine what your resource levels are and make decisions from that perspective. This is technically referred to as "state-dependent" reproduction.

This is how I have made my reproductive decisions. I have had neither sufficient financial resources nor familial support to feel confident about ensuring the best possible chance of success for my offspring. Too many bad years and too much uncertainty often lead to forgoing reproduction. Heck, even one bad year and an entire population can cease producing the next generation. We scientists call this the "bad years effect."

This was the case for the fat dormouse one fateful year in Germany. The fat dormouse is also known as the edible dormouse because it was historically considered a delicacy by the Romans and Etruscans, an extinct civilization from what we now know as the Tuscany region. The fat dormouse looks like a mouse that had its back end taken over by a squirrel. Unlike squirrels, they are nocturnal and squat in bird nests or tree hollows during the day. Although they aren't terribly social creatures, dormice in a given forested area will communicate with each other. When they aren't squeaking and snuffling about, they are leaving notes for each other using the scent glands on their feet.

Because they hibernate six months out of the year, they get busy making babies from June to August. Normally, not every dormouse reproduces every year, but in 1993 not a single dormouse in Central Germany reproduced.[13] What was the cause of this dismal output of new babies? The spring of 1993 was unusually dry and the dormouse's food supply of seeds, nuts, fruits, and berries was severely limited. Unable to fatten up properly, no adult was able to produce a baby dormouse. Unlike the dormice in this particular patch of forest that only experienced one bad year, I have had a lot of bad years! I suppose I could take comfort from a broader perspective across species that my decision was only natural. It seems like a reasonable approach and it's one that many people take.

We know from studies of twins that the cost to the mother is higher when resource availability is poor. Even without twins, simply having more children increases the chance of malnutrition of all the children. This is presumably because, as we see with other animals, there is an increase in the demand for resources that may not always be met. This raises an interesting question: Is it only the wealthy, or those with excess resources, producing the highest number of children? We might expect that a mother with higher social status and wealth would have more kids. After all, that is what we see in the dormouse and many other species that regulate their reproductive potential on the basis of how much food, shelter, and other resources are available.

By evaluating complete genealogical data cataloging three generations of Finns during the preindustrial age from 1709 to 1815, scientists were able to untangle this contradiction. Like in other parts of the world in the 1700s, health care had not yet become a part of mainstream life in Finland. This fact, combined with unpredictable crop yields, famines, unusual weather patterns, and an abundance of infectious diseases, makes this a fertile time period for investigating the relationship between family size, survival, and socioeconomic status.[14] What the scientists discovered was that women from lower and higher economic backgrounds tended to have the same number of children (approximately six on average), but that the cost to mothers was greater for those who were poor. Although among the wealthy and the poor there might have been an equal chance of their children dying from disease, mothers of lower economic status died younger. Other countries, such as Italy from the fifteenth to the eighteenth centuries, followed a trend we might expect: The wealthiest had more children.

Have things changed in contemporary society? Yes. By the twentieth century there was a switch and the poorest families began having the highest reproductive rates. Data from the United States Census Bureau in 2010 reveals that there is an almost twofold increase in the number of births among households reporting an annual income of less than $10,000 compared to $75,000. Why? One reason may be reduced access to education, health care, and contraception, all of which have strong inverse relationships with contemporary birthrates.[15] And while mortality rates were similar between the wealthy and the impoverished prior to the industrial revolution, this is no longer the case. The wealthy have better access to medical care, nutrition, and education, and their infant death rates are lower, so having more children as insurance has become obsolete for those of higher status. For those of lower economic status, however, low birth weights and infant mortality remain high.

If we think about economic status as a measure of environmental predictability—where the more resources you have, the more stable your environment is—then we can see that sometimes having more offspring in uncertain conditions is a viable strategy. If we consider quantity versus quality, where here quality does not imply the *intrinsic* value of a person, but rather his or her chances of survival, then sometimes hedging your bets by having more offspring might improve the odds that at least some will survive. This is similar to the strategy of the bigfin reef squid, albeit on a much smaller scale.

Sometimes, though, this approach is not adaptive, meaning it does not increase survival, where, again, survival in the evolutionary sense equals reproductive survival. Essentially, do your offspring go on to have more offspring than someone else? Great tit birds in boreal forests seem to be having a bit of trouble in this regard. Specifically, they consistently have more chicks than they can feed.

Usually bird parents have to be able to predict how much food will be available when they decide how many eggs to lay. Research shows that most birds are successful at figuring out how many eggs to lay and how many chicks they will be able to fledge. If conditions are terribly uncertain, the best bet is to reduce the number of eggs laid. In most places great tits do exactly this. They opportunistically have more chicks under better conditions and adjust by having fewer chicks under substandard circumstances.

Not in the highest latitudes of their range, though. Those great tits that are breeding up in the northern part of their range—Norway and Finland—seem to miss the mark. As a result, their breeding success is relatively poor compared to great tits living in other areas.[16] Are modern humans, like the great tits, reproducing maladaptively? We already know that individuals of lower economic status have more children, and those children have lower birth weights and higher infant mortality. The real key is whether there are differences in lifetime reproductive success that correspond to status and fertility.

Given that the changes in fertility are a relatively recent phenomenon, this remains unclear. What we do know is that—unlike the great tit, the bigfin reef fish, or the fat dormouse—we are on the precipice of overpopulation that, without restraint on individual family size, will impact all of us. If other species can regulate their population growth to be sustainable, then perhaps we can, too.

WILD LESSONS

* There are costs associated with having more than one child that go beyond the wallet, influencing everything from birth weight to IQ.
* How many children to have is a very personal decision. However, animals show that deciding based on resource availability and life conditions may improve success.
* Having more children than a parent can reliably support and successfully raise is maladaptive.

Playing Favorites

As I mentioned earlier in this chapter, transitioning from one child to two children introduces many variables, one of which is the opportunity for parents to favor one child over another. I come from a modest family size of two, and it has always been clear that my older brother was the favorite. My cousin and I commiserate on how her older brother is also the preferred child. Ironically, her mother (my aunt) and my mother both complain of how the sun rose and set over their older brother.

How was it that I came to unequivocally understand that my brother was the "golden child"? Well, other than being told how much better he was at just about everything, important resources were unequally distributed between us. He received more attention, more affection, more support in his education, and probably more food.

Was the obvious preference for one child over another a peculiarity of my family that has been passed on from one generation to the next? And was it a coincidence that the firstborns were the favorites? Did it have something to do with being male? Or being firstborn *and* male?

I can rule out any peculiarities of my family fairly quickly. Though not ubiquitous, it is unsettlingly common among humans for parents to favor one child over another. The estimates range from two-thirds to three-quarters of parents exhibiting a preference for one child. Psychologists call this "differential treatment" of offspring.

For chinstrap penguins, we call this normal. Found off the coast of the South Pacific and Antarctica, chinstraps owe their name to the delicate strip of darkened feathers that, when combined with their black heads, makes it look like they are wearing a helmet. Roy and Silo were two famous chinstrap penguins that called New York City's Central Park Zoo home.

Penguins are typically spectacular parents and these two males were no exception. Roy and Silo were so interested in having a chick that they turned their parental affections toward a rock. Yes, a rock. They were devoted to their rock until a keeper provided them with a real egg to incubate. (As an aside, same-sex partnerships are fairly common in other animals, particularly birds. In those cases where they raise offspring together, they do as good a job (if not better) than mixed-sex parents. That should help put to rest any notions about the parenting abilities of same-sex partners.)

Now, back to those dastardly penguins and their unequal treatment of their kids. It's a good thing the folks at the zoo only gave them one egg. Chinstrap penguins, and other species of penguin, often make their kids chase them for food, especially if they have two at the same time. Perhaps baby penguins are annoying and the parents are attempting to tire them out? Or maybe they simply can't cope with begging chicks and are trying to run off? Those would be the most innocuous, even humorous explanations.

Unfortunately, though, these explanations are just wishful thinking on my part. When there are two chicks to feed, the parents run away because of favoritism—they're trying to feed one chick over another. The way that chinstrap penguins decide whom to feed more is with a race. The chick that can't run fast enough after the parent to get enough food ultimately doesn't make it.[17] This is how they regulate who gets what, and when.

In doing so, the parents are ensuring unequal access to food; the strongest and fastest eats most. In some ways, people who come from large families can relate to the scramble for food. As soon as the food is set down for dinner, it's a mad dash to see who can get the most first. Unless parents intervene, one or more children may find themselves on the short end of the stick. Incidentally, experiencing food shortage as a child impacts one's relationship with food as an adult. Have you ever noticed someone who inhales the food on the table and looks around, suspicious that someone else might be taking more than a fair share? Or someone who rushes to eat first and takes the most? You ever notice that?

The type of favoritism exhibited by chinstrap penguins is rather passive in the sense that the parents aren't deliberately serving one child more food than another, but they aren't guaranteeing equal access, either. What about more directed discrimination in favor of one child versus another? My cat family illustrates a very purposeful and deliberate prejudice in this regard. I have a family of cats, a mama and her two, now thirteen-year-old, kittens. I have a love-hate relationship with the mother, Midnight, precisely because of her disparate treatment of her offspring.

After about three to four weeks of nursing both her daughter, Peanut, and her son, Buttons, Midnight began attacking her daughter every time she tried to nurse. She did not, however, attack her son. She continued to generously nurse him for several weeks after she stopped nursing Peanut. She also played with and groomed Buttons much more, while routinely rebuffing sweet Peanut. Peanut is not so sweet anymore. She is quite neurotic about food, gobbling it up so quickly that she frequently gets sick, and she doesn't seem to have learned how to play properly. It makes dealing with her a challenge, to say the least. To this day, her mother will only play with Buttons, romping and chasing him around, the two pouncing friskily on each other. I often catch Peanut gazing wistfully at them and I imagine she wishes she were part of the fun.

We arrive at an interesting juncture in the discussion of child preference. Do parents exhibit a preference for one gender over another? There is research to support the notion that mothers frequently prefer their sons. This may be subtle or, as in the case of some, like Midnight, an unrestrained bias. But in humans this is highly variable and dependent, in part, on cultural norms. For instance, because of the high value placed on sons in India, Indian women tend to wean daughters sooner than sons, especially if they do not have a son yet and want to get pregnant again. Perhaps, relatedly, that is why there is a higher mortality rate among girls in India.[18]

Wandering albatross female chicks also get less. Like most humans, wandering albatrosses have one chick at a time, which allows a straightforward peek into how they treat sons and daughters differently. Albatrosses got the nickname "gooneys," in the foolish sense, because they are remarkably docile on land. They were revered by seamen, who were reluctant to kill an albatross for fear it would bring misfortune upon them. The wandering albatross has a circumpolar distribution and has an astounding ten-foot average wingspan. Given that all birds have at least some portion of bones that are hollow to reduce the energy expenditure of flying, it is nothing short of remarkable that adult wandering albatrosses weigh upward of fifteen to twenty pounds.

To get this big takes time, and chicks need many months to get their plumage and years to become fully breeding adults. They also require a lot of food from their parents. Like human boys, boy albatrosses grow larger and thus require more food (imagine a hungry male teenager). Therefore, wouldn't it make sense to produce relatively "cheaper" females?

Yet when scientists took a snapshot look at one wandering albatross colony on Île de la Possession in the Crozet archipelago, this is not what they found. Possession Island is the only inhabited island in this subantarctic archipelago and is manned only by research staff. Interestingly, these islands are French territories and have been classified as national parks since 1938, protecting wildlife. The French literally traveled to the end of the earth when explorer Marc-Joseph Marion du Fresne discovered these islands in 1772. His demise came shortly thereafter when he, along with many French sailors, were killed and eaten by the Māori in New Zealand. But back to wandering albatrosses.

When the researchers looked closely at the sex ratios, feeding rates, and survivorship of chicks and their parents, they discovered a few interesting things. First, the sex ratio, or number of male to female chicks born, was ever so slightly skewed toward boy chicks. Male chicks were more likely to survive, grew faster, and weighed more by the time they were fledged. Why? They were fed more. Parents of boy chicks were able to synchronize their feeding schedules so that the time between meals was less uneven. Male parents really stepped up and delivered more food to their sons than to their daughters.[19]

When it comes to humans, it would be an oversimplification to suggest that a patriarchal society or culture alone drives differential investment in males versus females. We have already seen that environmental conditions can influence how many children we have, but is a biological component that has its roots in natural selection unconsciously impacting our behavior? Remember Robert Trivers from our earlier discussion of parent-offspring conflict? He and another colleague, Dan Willard, postulated what is called the Trivers-Willard hypothesis.[20]

The heart of their argument suggested that for some species, when conditions are excellent, females should preferentially have and invest in sons over daughters. When conditions are reversed, females should similarly switch and divert their efforts toward raising daughters. This ultimately means that mothers in outstanding conditions are expected to either (a) produce more sons or (b) direct resources differentially toward sons. What is the basis for this reasoning?

It all comes down to sex—specifically, the greater amount of sex that healthy, strong sons get to have when they grow up. Let me explain. If a mother is in excellent condition, she is more likely to have a baby that is born in excellent condition (e.g., with a healthy birth weight). Presumably, since she herself is very healthy, she will be able to provide enough food (determined by lactation amount and duration) for her new baby. This, in turn, means the baby will have a greater chance of survival into adulthood, because being on the receiving end of the proverbial silver spoon has lifelong benefits. In species where some males have the potential to produce disproportionately more offspring than other males and all females, it pays reproductive success dividends for mothers to invest more in sons when they have the best chance of producing "sexy" sons that go on to father many offspring. If conditions are suboptimal, mothers may as well invest more in daughters, since females are always limited in the number of offspring they can produce (relative to males).

Does this hypothesis hold up? It does for the tammar, or darma, wallaby. Endemic to parts of Australia, it is one of the smallest wallabies. However, don't be fooled by its modest size or baffled expression; the tammar wallaby is not a wimp. Years ago a friend of mine working as a zookeeper in Florida got into a kerfuffle with a wallaby. He bears a thick, angry red scar from this encounter and, needless to say, his respect for wallabies went way up.

Like other marsupials, tammar wallaby mothers carry their joeys, or babies, in their pouch. In keeping with the Trivers-Willard hypothesis, research shows that when mother wallabies are in good condition, they are more likely to have sons and nurse them more.[21] Does this play out in our species? Maybe. The science is equivocal.

Let's move away from feeding and toward something more meaningful to humans today: education. In theory, as your education level increases, the greater your probability of economic, social, and reproductive success. This may be more pronounced if you are male, although it improves marital prospects for both sexes.[22] Using data from 2000 to 2010, and calculating a socioeconomic index (SEI) composed of a combined average of levels of education, income, and social prestige measures, it was discovered that, on the basis of a father's SEI alone, it was possible to predict which sex would be on the receiving end of higher parental investment. A higher SEI of the father led to greater parental investment in boys, while a lower SEI resulted in a higher investment in girls.[23] The study only used the father's SEI because, rightly or not, it was assumed that a woman paired with a male of higher status would automatically be in better condition.

I started this section by sharing the intense favoritism that pervaded my family dynamics. Certainly my brother's education was supported in a way that mine was not. He got to go away to college and went on to become a doctor. I slaved away waitressing to pay my way through undergraduate school, and although I did get a PhD, it took me a full six years longer than average. As I have mentioned, I am not alone in this experience. Through our rose-colored glasses we seem to be unwilling to acknowledge the discrepancies between our ideal vision of parent-child relationships and reality. The research abundantly demonstrates how mothers, in particular, are closer to and provide more support to one or more of their children over others— more often than not. Some of this may be due to life circumstances, including marital stress or divorce.[24] Stressful conditions can alter how adequately a mom feels she can meet the needs of all her children, thus predisposing her to be less available and giving to some and not others.

In my family, maybe it was because my brother was the firstborn, maybe it was because he was male, or maybe it was because of the divorces—but there is another possible source of favoritism that we cannot ignore: personality and temperament. In my family, my brother's personality traits and disposition were more similar to my mother's, or at least preferred by her. This alone can trigger differential treatment. Let's face it, I had a lot of strikes against me. I looked and acted more like my father, who by the time I was six had abandoned his responsibilities as a parent. I was not a little princess by any stretch of the imagination. I played in *dirt*, refused all dresses, insisted on climbing trees, and had a menagerie of pets—of which mice were my hands-down favorite. None of these characteristics were considered endearing in my family.

However, not every instance of differential treatment means something nefarious. Newborns by definition need more attention than older children. A highly anxious child may require additional emotional support. An aggressive or bold child may require stronger boundaries. The key is communication. Parents can, and should, explain to their children any differences in treatment and balance the scales in other ways to diminish the risk of serious problems emerging in their "disfavored" offspring. Aside from causing psychological problems, including depression, low self-esteem, and reduced overall success, favoritism also leads to sibling hostility and resentment. Arbitrary favoritism is fertile ground for sibling rivalry, an already common state of affairs. Only by recognizing our tendency to play favorites can parents make an effort to level the playing field.

WILD LESSONS

+ As family size grows, the risk of differential treatment also goes up.

continues . . .

- Many an albatross, penguin, and human exhibit a preference for an older child, a male child, or one that is more similar to themselves.
- Stressful life conditions can increase the probability of treating children differently.
- By recognizing and correcting imbalances and treatment, parents can avoid the serious impact that favoritism has on their children.

Sibling Rivalry

When I was eleven, my brother set me on fire. To be more precise, he set my hair on fire. Although it sounds somewhat humorous now, given that I survived, it remains true that sibling rivalry is no laughing matter and makes up a significant portion of the violence in family homes. And yet, even with the stitches, burns, confinement in small closets, torturous tickling, and relentless mistreatment, I still had it better than a sand tiger shark.

Sharks in general are remarkable. Scientists are now discovering that sharks are not these mindless machines out to ruin our seaside getaways. They are individuals, with personalities, preferences, and feelings. That being said, sand tiger sharks engage in a death battle with their potential siblings before they are even born. Female sand tiger sharks have a double uterus, or two uteri. So it might seem obvious that they would therefore have just two babies. Ultimately she does, but it doesn't start out this way. In each uterus there may be dozens of embryos. When the embryos hatch, they have teeth. These teeth wouldn't be strong enough to pierce your skin if your hand mysteriously found its way into a sand tiger shark's uterus, but for the sand tiger shark hatchling that emerges first, they are sharp enough to slash and eat all of its yet-unborn brothers and sisters. This happens in each uterus. What's more, after the pair of greedy "firstborns" eat their siblings, they move on to chomping on mama's extra stash of eggs.

The baby sand tiger sharks finally emerge chunky and healthy, off to a good start in life.[25]

Sand tiger sharks are not alone. Although intrauterine consumption of one's siblings is relatively rare, killing one's sibling is commonplace in many species, especially birds. Where there is viciousness among siblings, scientists usually blame the parents. That is because the parents often do not intervene, and indeed may rely on sibling aggression to reduce brood or litter size. That is definitely the case with the masked booby.[26]

The masked booby, one of six species of booby, is notorious for its penchant for "siblicide," the formal term for killing a sibling. Making its home in the tropics, the masked booby is a magnificent seabird and an outstanding hunter that dives at incredible speeds to catch a meal of fish, squid, or octopus. Siblicide in this species is necessary for survival and facilitated by the parents, who build a shallow nest to assist the first hatched chick in kicking its sibling out. The chick does so by using its beak to grab its sibling by the neck, wing, or other body part and fling its brother or sister out. Why do masked boobies do this? In this species, it's all about insurance. Hatching their eggs seems to be a challenge for these boobies, with eggs hatching less than two thirds of the time. To avoid complete nest failure, the female lays two eggs, even though she and her partner can only afford to raise one.

Among mammals, including humans, killing one's sibling is far less common. Within human families, rates are low (less than 3 percent), and where siblicide does occur, it is frequently between brothers.[27] Of course, now we're getting into the most extreme side of sibling rivalry, far removed from the day-to-day challenges many human parents face addressing conflict among their children. The first thing to understand is that, depending on personality or other factors, it is rather reasonable for siblings to compete. Each offspring is primarily concerned with its own survival. From a purely selfish perspective, it is beneficial for one child to attempt to sequester as much as he or she can from the parents, even to the detriment of his or her sibling. If within a family this approach is being undertaken by each child, this sets the stage for sibling competition. Children are hypersensitive to the amount of attention and resources they are receiving relative to others, as well as to the imposition of boundaries. Fairness is quintessential to their lives when additional kids start showing up in the family, and sibling aggression should not go unchecked.

Although we will discuss the cooperative aspects of meerkat families in Chapter 8, suffice it to say that meerkat siblings routinely engage in aggression that can be fierce, though it rarely results in serious injury and is never lethal—similar to the day-to-day conflict happening in human households everywhere. Meerkats are carnivores that belong to the mongoose family. They are adorable but their disposition leaves much to be desired. They are scrappy, to say the least, and their training begins early on. A single dominant female will have several litters a year, which is possible because other meerkats act as helpers. When it comes to getting fed, the best way a young meerkat can secure food from a helper is to physically be the closest. To do that they are not above snapping or lunging at a brother or sister who tries to sneak past to get to a helper.[28] Having a regular helper gives a young, growing meerkat an edge. The lack of injury that results from these scuffles means that the cost of being aggressive to its brother or sister is low. One can see, then, that there is a lot to gain and relatively little to lose from such sibling aggression. That may be why it is so ubiquitous during development.

However, conflict among siblings can make things more difficult for parents. How do animal parents reduce hostility and jealousy among their children? Galápagos fur seal mothers typically have little patience for older siblings harassing younger ones. In this species, about one quarter of all pups are born while an older sibling is still around and being nursed. You'll recall from above that often the second child weighs less. Well, this holds true for pups that come into the world under these circumstances, which creates an additional disadvantage for them. Their not-so-loving brothers or sisters regularly chase, bite, or otherwise attack them, doing everything in their power to discourage the newborn pups from feeding. Luckily, mothers aggressively defend their newborns from their older pups with a combination of threats, and biting if necessary, to discourage the fighting. Most of the time this is successful and the older ones accept that the time has come to grow up—unless it is a very bad year and times are tough, in which case older siblings don't take too kindly to their mothers shoving them off the nipple and resist. If the older siblings challenge long enough, moms may begin to neglect the newborn pups, who will die of starvation.[29]

Similarly, during childhood and adolescence, human sibling relationships tread a fine line between conflict and cooperation. One big difference between humans and other animals in the arena of sibling conflict is that the most commonly cited source of struggle between human siblings is the sharing of personal possessions, followed by power, aggression, outside friends, and violation of family rules. It would seem that Freud got it wrong and siblings are less concerned with what is going on between the parents and siblings than what is going on between the siblings themselves. As a result, siblings rarely argue over parental affection. So how do siblings resolve their disputes? They don't. More often than not parental intervention is the principal way conflicts are resolved.[30]

How parents respond, mediate, or intervene can greatly shape the nature and trajectory of individual development and sibling bonds. Despite some expert's suggestions, nonintervention is not the optimal choice, particularly in aggressive interactions. If sibling conflict were truly about parental attention, then perhaps nonintervention would be viable, but as mentioned above, this is rarely the source of angst between siblings. More importantly, without any parental interference, coercive and abusive (psychological or physical) sibling conflict unsurprisingly leads to problems for the individual under attack.[31] Possibly the only thing worse than not intervening is actually sanctioning physical aggression from one sibling toward another.

How do parents typically handle quarrels among their children? Facilitated negotiation and compromise is the most common approach, followed by doing nothing, and lastly, actually supporting aggression between their kids. When children are closer in age, it seems that endorsing physical fighting between kids is even more common.[32] But we already know that failing to intervene isn't an effective strategy for mediating arguments among siblings, and now setting aside the disturbing reality that some parents tolerate and permit physical aggression as a viable tactic, let's turn our attention to coaching a peaceful negotiation. Negotiation can be considered a form of cooperation, and although we need to be schooled by our parents in this process, barn owl chicks take to it much more readily.

Barn owls are one of the most widespread birds, found everywhere except in extreme environments, north of the Himalayas, and Indonesia. They are simply stunning, with hauntingly white heart-shaped faces. And, despite their name, they don't live in barns. (In fact, they actually have a variety of names, including ghost owl, white owl, death owl, and hobgoblin owl, to name a few.) They like small roosting and nesting sites and can be found using holes in trees, cracks in cliffs, abandoned buildings, and even chimneys.

Just like other birds, barn owl chicks beg for food from their parents, but some recent research sheds light on how sibling barn owl chicks may be calling to influence each other more than their parents. In essence, if siblings can communicate their needs to each other, the hungrier one can do most of the calling, while the other one can conserve energy. All that yapping isn't cost-free. Thus, the pair would be cooperating and negotiating who got what when based on degree of need. Early evidence reveals that barn owl chicks communicate with each other *before* their parents get back to the nest with food. What is interesting is that it wouldn't be hard for one sibling to completely dominate another. Barn owlets hatch asynchronously such that there are obvious size differences between all the brothers and sisters in a nest. Coupled with the fact that the food cannot be split up among each sibling, since the parents feed them tiny critters such as shrews, and the fact that baby barn owlets can eat their body weight in food every night, it is quite surprising that there is more conciliation and cooperation than straight-up competition. Here's how this cooperation actually works: The hungrier an owlet is, the longer and more frequent its calls compared with its less-hungry siblings. Each owlet compares how hungry it is to the calling rate of its sibling and adjusts its begging rate based on the chance it will succeed in getting food from its mom or dad relative to its much hungrier sibling. Gosh, that is complicated!

Basically, the hungriest owlet gets everyone else to keep the volume down so it can get the food. And the siblings comply. From a cost-benefit perspective, it's rather like saying, *Boy, it seems like my brother is really hungry. I am not that hungry, so instead of spending my energy and all that effort calling loudly once my parents get back to the nest, I think I will just chill for now.* Unlike some human siblings, barn owlets don't try to take more than they need just because they can. They negotiate among themselves.[33] As a parent, if you've got more than one child, it can be helpful to be able to discern real need from a sibling trying to control access to resources, and allocate accordingly.

Although some anthropologists suggest that humans are cooperative right out of the gate, I would say some people are and others need to be taught or coaxed into sharing, being helpful, and cooperating with others. This begins at home. After all, how siblings treat each other is their first lesson in being social and part of a group. Therefore, perhaps the single most important way that parents can address and mitigate sibling rivalry and conflict when and if it arises is to foster a greater sense of cooperation within the family unit.

My good friend Paul, who had those reccurring obsessive thoughts of dropping his first child, a daughter, down the stairs, described to me what it was like raising his two subsequent children, boys who are only nineteen months apart. They are still very young and thoroughly enjoy each other's company—as long as they both like what they are doing. Most spats involve jockeying for position or just arise from boredom. As a fellow evolutionary biologist studying the evolution of cooperation, he has a unique perspective as a parent. And he's come to believe that families that create a sense of common purpose will generally have a lower level of conflict. By demonstrating and teaching his children that the family unit as a whole benefits from the collective and coordinated action of each individual, he reduces competition between his children. What is important is that each individual also feels that he or she benefits from being part of the family.

The other perspective he shared, and one which I wholeheartedly agree with, is that one should never underestimate the importance of treating one's children fairly and consistently, without allowing them to exert too much in terms of individual demands. A delicate balance, for sure, but if one fails to do so, and instead indulges each individual child's request, the stage is set for siblings to be locked in a near-constant battle to get what they want. His warning: Do so at your own peril.

We as Americans in particular place the individual above all else, and the concern for self *at the expense of others* breeds discontent, competition, and discord. We see this manifest in relationships and within families, and by extension in society at large. If you want your siblings to treat each other well, don't let any individual child gain an unfair advantage over others in the family. True cooperation will emerge from fairness, equality, and benefits to the individual *and* the collective. We like to say life isn't fair, and it certainly isn't, but we are all incredibly interested in being treated fairly. We are born scorekeepers, some more than others. You can try to eliminate this trait in your kids, but they will keep score anyway and grow up to resent their parents, rightfully so. Practicing fairness emerges as a way to foster long-term cooperation, and it is unreasonable to expect anyone to cooperate if someone else in the family accrues a greater benefit.

At this point you might be thinking I am portraying a dark picture of sibling relationships, so let me turn the corner and move on to the benefits of having brothers or sisters. Obviously, not all siblings will experience conflict. One of the great things about having a brother or a sister is having someone to watch your back, to help raise you, and to work with so that you can both succeed.

Cheetahs know all about how forming an alliance with a sibling can increase the chances of survival. I met a cheetah named King George once. He was new to the Miami Metro Zoo (now named Zoo Miami) at a time when zoos were beginning to use cheetahs as ambassadors to garner public interest. He has since died, but when I met him, he was a young, stunning one-year-old feline. I have always had a fascination with predators, and was honored to have the opportunity to spend a little time in his company.

Cheetahs are stealthy, fast, but fragile hunters at the mercy of larger, fiercer competitors such as lions, hyenas, and leopards. That is most likely why, after leaving their mother, young cheetah siblings, both brothers and sisters, stick together for an average of six months. After that, a sister takes off so that she will be free to mate, but brothers remain together throughout their lives.[34] This is particularly interesting because, unlike male lions that also form coalitions, male cheetahs are not getting the benefit that lions do: access to more females. That is because, unlike lions, female cheetahs are solitary. Despite this, on the plains of the Serengeti almost 60 percent of male cheetahs are part of a group, either a pair, trio, or rarely, in groups of four. Why? Adult male cheetahs that travel alone are almost never able to secure a territory and thus are less likely to live as long.[35] It pays to stay with family, at least for the boys.

Even though my friend Alma only had one child, she comes from a family of four siblings, so she knows well one of the greatest benefits of having siblings: having backup. Often, among close siblings there is a rule: I can pick on you, beat you up, or otherwise make your life difficult, but if anyone outside of the family does it—watch out!

WILD LESSONS

* Sibling aggression can be a serious problem in some human families, though it rarely reaches the levels seen in sand sharks. Don't ignore violence among siblings.
* Contrary to popular belief, siblings are rarely competing for parental affection.
* Barn owl siblings cooperate and negotiate among each other. Teach children how to navigate their relationships fairly.
* Whether it is work or rewards, divide things fairly among siblings, and be prepared to offer a reasonable explanation for why it is fair should one child have a gripe.
* When siblings can cooperate and be there for each other, they will benefit from always having backup when life gets rough.

CHAPTER 7

They Grow Up So Fast

A saying that resonates with many parents is, *They grow up so fast!* This seems to apply to the early years, though. Later, this familiar phrase is replaced with, *Will they ever just grow up and move out?* And then when they do, for many parents, there is a moment (or more) of sadness as they watch their child go off to college, or get her own apartment, or move in with her boyfriend. As Alma explained, when she arrived at that moment when she had accomplished what she had promised her newborn infant she would do—to love her, teach her, and help her become the very best version of who she could be in this world—she simultaneously felt a profound sense of pride for the woman her daughter had become and a profound sense of loss over the mother that her daughter no longer needed Alma to be.

Parenting isn't just about providing children with their material needs; it is also about teaching them how to be in this world as humans, just as rhinoceros mothers must teach their offspring how to be rhinoceroses, or blue-footed boobies must impart everything their chicks need to know to be successful boobies. As all youngsters grow and develop, the lessons become more serious and complex, and very few of these lessons are completely innate, especially if you are a social species, like us. Navigating the world requires traversing all the stages of development effectively, and parents are the compasses guiding their children through the sometimes rough waters of childhood into adulthood. There are the mechanical elements that kids need to learn, such as how to hold a fork or a cup and how to tie their shoes. There are also important social skills they must begin to develop, such as sharing, making friends, and resolving conflicts. And so much more. As I considered all of this, I began to wonder, how do parents decide when a child is ready for the next steps on the path to independence? Does the pace depend on the child, on the phase, both? Naturally, this led me to be curious about how other species make these decisions and deal with the challenges that come with each new step toward independence.

Feed Yourself!

Possibly one of the most universal first steps toward independence across all species is learning how to feed yourself. In Chapter 5, I talked about the average duration of nursing in humans and other species of mammals, but being weaned is not the same as feeding yourself, as any parent whose kitchen floor is splattered with puréed sweet potatoes can attest to. First of all, offspring need to be taught *what* to eat. This is not a trivial matter, especially for honey badgers. Honey badgers are like the Sylvester Stallones or Arnold Schwarzeneggers of the animal kingdom and belong to the same group that includes weasels, otters, and wolverines. Very little detailed information is known about honey badgers, but for those that live in the Kalahari Desert, researchers have been able to glean a bit about their feeding ecology and behavior.

For me, the Kalahari has long been a magical, mystical land that I hope to visit. When I was an undergraduate student studying psychobiology in Florida, I stumbled on the wonderful books by Mark and Delia Owens, among them *Cry of the Kalahari*, a tale of their adventures as scientists in this remote, wildlife-filled, harsh wilderness. Honey badgers make a living there by having a varied diet that includes some downright dangerous items on the menu, including wild cats, scorpions, poisonous snakes (Cape cobra), and, of course, bee larvae. With almost sixty species to consume,[1] a mother honey badger has a lot of work to do. Teaching young honey badgers how to safely subdue and kill hazardous foods like cobras or carefully handle a beehive so as to not be stung to death by angry bees is crucial. For honey badger cubs, until they are three months old their only forays out of the den are via their mother's mouth as she transfers them from one site to another. As with human children, there is a transitional period where their mother introduces them to different foods, and it takes almost a full year for them to be able to feed themselves independently.[2]

Like honey badgers, humans have a very diverse diet, and our children learn what to eat from us. This is also true of many birds. In a clever experiment, scientists swapped eggs between blue tits and great tits. Normally, these two species of passerine, or perching birds, coexist happily side by side and eat in different parts of the forest. Blue tits like to be higher up from the ground than great tits and also are better suited to looking for food in buds and twigs. The question scientists were interested in was whether or not the chicks learned from their parents and what would happen if they were raised by another species. Even the researchers were surprised that the cross-fostered chicks picked up the feeding habits of the parents who raised them.[3] Simply being repeatedly exposed to feeding higher or lower off the ground, despite the natural advantages there might be in sticking with what routinely works for their species, is all it took to make the switch.

What does this have to do with us? One of the major challenges for up to 50 percent of parents is dealing with "picky eaters." There are striking differences between the nutrient intakes of children who are picky versus those who are not, including lower amounts of vitamin E, all the B vitamins, and iron in the picky group. It is well established that a child's willingness to accept a new food is directly linked to watching his or her parents and siblings consume those foods. However, most parents stop trying to give a child a new food after a mere three times, despite the research suggesting eight to fifteen times is optimal.[4-5] As is the case for blue and great tits and a myriad of other species, which foods human parents expose their offspring to is the single biggest predictor of diet later in life.[6] And a broad diet is crucial to chidlren's development and health as they age. So keep at it, because repeated exposure is the only way to go to ensure your kids learn what to eat.

WILD LESSONS

* Getting kids to try new foods seems a lot harder for humans than for other animals.
* Orangutan mothers partially chew up their food before passing some to their offspring to try.
* As soon as children are transitioning to solid food, support their curiosity and expose them to as many safe foods as possible.
* If your kid rejects a food item, don't give up! Follow a blue tit approach and keep trying—but don't force it.

Nighty Night!

Another big step toward independence is sleeping alone. Even though, as we saw in Chapter 5, co-sleeping is ubiquitous around the world, and there are clear benefits to sleeping with one's parents for infants under one year old, sooner or later, a child must learn to sleep on his or her own. The big question is when? Is there a "right" age? Does it depend on the child, the personality, the temperament, or other factors?

For some species, individuals may sleep with their parents or other family members throughout life. Before launching into my dissertation research on prairie dogs, I had the opportunity to travel to my advisor's field site near Iguazú Falls, Argentina. He was working on spatial memory and foraging patterns of the tufted capuchin monkeys and I was contemplating a study of predation risk. My first encounter with the troop of monkeys rivaled my encounter with the thieving vervet in Kruger National Park in that I was schooled on the hierarchy by the lead male, Gerdame. The social organization of tufted capuchin is such that the group is composed of several males, related females and their offspring, and a few peripheral males, all led by the alpha male. Gerdame took his leadership role *very* seriously and upon seeing me, the newest member of the field crew following his group, he dashed down the trunk of an enormous pindó palm tree, stared me straight in the face, and screamed at me. I was utterly convinced of his superiority and immediately looked down and backed away . . . slowly.

When it comes to sleeping, tufted capuchins sleep together in the same general area, utilizing certain trees that are large enough to keep the group in close proximity to one another. Infants and juveniles up to one and a half years old always sleep with their mothers. After that age they still sleep in close proximity to someone—sleeping alone very rarely occurs (only approximately 10 percent of the time).[7] Capuchin monkeys are known to take a long time to develop relative to their body size and reach the equivalent of the "teenager" stage around five years old, yet they reach the "no longer sleeping with mom phase" a few years earlier, probably because their mom may now be preoccupied and investing in a new infant. Whether or not juvenile tufted capuchins protest being displaced by a new brother or sister is unknown, but we do know that human children may protest.

Why do some kids embrace having their own bed to sleep in without so much as a hint of discomfort, while others resist as if their very lives depended on it? One answer could be that this is a manifestation of parent-offspring conflict, where some children want a certain high level of parental attention, care, and affection to continue beyond what the parent is capable of or willing to provide. In today's modern society where sleep is a precious commodity for adults, it is no wonder that sleep behavior is fertile ground for this conflict to take place. For parents, the reason for transitioning a child to sleeping alone may be as simple as needing a full night of sleep. If they are unable to achieve this while sharing a bed or a room with offspring or having to make frequent trips to a child's room to provide comfort, they may terminate this behavior sooner than a child is willing to accept. Some kids would be more than happy to have their parents rub their backs all night while they dream of bunnies and rainbows. This discrepancy between what the child wants and what the parent is willing to give becomes the source of parental struggle.

Certainly as adults we all benefit from being able to sleep alone, soothe ourselves, and go back to sleep. Naturally, it is also important for children to master this. What is less clear is which age, developmentally, is the logical time for sleep training, or if the decision is largely arbitrary.

Paul decided to begin sleep-training his youngest son at one and a half years. Until this time, the child slept in the same room with his parents. This decision to move his son to a separate room was based primarily on Paul's desire for privacy, for a place to be alone and talk with his partner, not on whether his child was ready for this transition. His older son, age four, still struggled with sleeping on his own, so Paul also decided to hire a sleep consultant to help. The sleep consultant's job was to help Paul and his wife implement a plan for this changeover to take place, but according to Paul, the real job was to give them permission to make this transition and cope with the emotional fallout that might ensue. For Paul's younger son, it was tumultuous, to say the least. There was screaming and crying, and finally such intense tantrums that vomiting resulted. Paul felt strongly about continuing because, as he put it, his son needed to learn how to soothe himself and put himself back to sleep. Gradually, over a few weeks, there was success, and his younger son surpassed his older one is his ability to sleep on his own.

Do we see other species struggle with sleep issues? I'm not certain that anyone has examined this matter empirically, but there is anecdotal evidence that failing to enforce boundaries and help one's offspring make this leap into independence can have disastrous effects. Flo and Flint, a mother-and-son pair of chimpanzees, immediately come to mind. Flo was part of Jane Goodall's original study on the chimpanzees of Gombe and arguably an extraordinary mother. However, by the time she had Flint, she was advanced in age. Flint was a rambunctious and demanding chimpanzee. When he was about five or so, Flo had another son, Flame. Unlike with her previous children, Flo had been unable to get Flint completely weaned, and he was reluctant to sleep in his own nest. Flame passed away when he was six months old, and after that, Flo seemed to give up entirely on setting strong boundaries with Flint. When Flo died in 1972 at approximately forty-eight years old, Flint halfheartedly built a nest, lay in it, and died a few weeks later. He simply wasn't able to cope without his mother despite being old enough to survive and thrive without her.

For some kids, it's less about simply wanting parental attention and companionship at night and more about need, in which case, they are genuinely not ready to make this transition to independence. They may be more sensitive, frighten more easily, or merely require more time than what is convenient or desirable for parents.

If you think about it, sleep is a time when all of us are vulnerable, a fact that I didn't fully grasp until I was in graduate school, where the evolution of sleep was a frequently discussed topic. The amount of sleep needed by any particular species varies widely, but everything that is alive spends some period of time at rest. We, like all other mammals, sleep. An armadillo rivals most cats in the amount of sleep it needs—twenty hours—while a donkey can get away with a mere three. Depending on where you are in the food chain, sleeping is more or less risky, and your ability to protect yourself is completely compromised. That is why most social animals that face any kind of threat sleep *together*, including most humans. It also may be why children, who experience this primal fear more directly, are so susceptible to becoming irrationally frightened. Shadows become bogey monsters and nefarious creatures seem to be lurking under the bed.

Part of this is also due to the massive capacity for fantasy that children have, which can sometimes lead to groundless fears. Until children are about six or seven years old, pretend play is incredibly important to their development. It facilitates the development of a suite of cognitive skills, as well as the theory of mind, which is the ability to attribute states of being to oneself and others—to put oneself in the shoes of another. Over time children learn to integrate feelings and emotions with reasoning.[8] The cost? When combined with a natural disinclination and fear of sleeping alone, childhood fantasies can make bedtime downright terrifying!

Experience can also alter or temper a child's willingness to sleep alone. When I was about nine years old, just before my Oma moved back to Brazil, someone broke into our house. My mother's new husband always insisted on keeping the sliding glass doors open and, unsurprisingly, it was easy for a thief to come into our house. Oma confronted the man, asking "What do you want?" and thankfully he took off. There was not a lot of uproar about it and the sliding glass doors, much to my dismay, remained open.

About a year later, a set of twins I knew at school were not as fortunate as my family. Because I was so young, the details were never fully made clear to me, but what I do recall is that an announcement was made at school that their mother was murdered. The whispers suggested that she was killed at night . . . while they slept . . . down the hall. I remember thinking, *Wait. You can sleep through someone killing another person down the hall? How can you protect yourself if you don't wake up?* My sense of safety while sleeping was shattered. I didn't sleep properly at night for another three to four years. With Oma gone, I had no one to climb into bed with when I became frightened. And because the open doors at night were now far more ominous to me, I would stay awake until everyone in the house had gone to sleep and quietly close them. After repeatedly being punished for doing so, I solved the problem by simply staying awake until the sun came up. Of course this created an altogether different type of problem: Sleepwalking, usually caused by sleep deprivation, was a common occurrence for me. In some sense this was a reversal of a state of independence I had achieved because, after those events, I became less confident and needed to rebuild the skill and "risk" sleeping alone again. Without parents patiently taking steps to support and guide me in this process, it took much longer than necessary.

WILD LESSONS

* Sleeping alone is a big deal. Most other social species sleep in close physical proximity to each other because one is most vulnerable when one is asleep, and at some level we all feel this.

* Young animals almost always sleep with their parents, but the duration varies widely.

* Your child may not be ready to sleep alone when you want him or her to. Regardless, sleep routines can help your child gain the confidence to make this transition.

* Some kids may readily embrace sleeping alone with little resistance, while others take longer. Decide if it's truly necessary to press the matter or if you are setting an arbitrary deadline.

* Children have a wonderful capacity for pretend play, and it is a natural and normal part of their development to have a rich fantasy life. Of course, this can lead to frightening images and scenarios involving sleep. Patiently and gently helping them to differentiate between fantasy and reality will not only assist them in their cognitive development, but will also help to slay those imaginary dragons.

* Any kind of traumatic event, mild or severe, can result in sleep problems for children. If your child, who previously slept alone, reverses and begins wanting to sleep with an adult, explore with him or her what is happening and take time to reestablish sleeping boundaries and behaviors.

Building Confidence

That brings us to helping our children reach those milestones and coping with negative events that may result in a regression. For humans and other animals, a big part of parenting is teaching kids how to be the best expression of who they are and equipping them to succeed in life. This involves walking the tightrope between imparting information about all the potential dangers out there while simultaneously encouraging children to take risks. It's a tricky balance, to be sure. We baby-proof our homes and try to warn children about all the hazards of the world: *Don't touch the stove. Look both ways before crossing the street. Avoid strangers.* The lessons are endless. We do this to give our kids a head start in life, since we cannot possibly protect them from every imaginable risk.

Young zebra finches get a few lessons too, although their schooling starts before they even hatch. Zebra finches are not just what we scientists call a model organism—a species used extensively to study a suite of topics, especially their flamboyant singing—but they are also model parents setting their chicks up for success. Before their precious babies hatch, zebra finch parents are preparing them to deal with the challenges ahead. What are zebra finch parents singing about? The future!

By talking to their kids about the weather, using a highly specific call given only when temperatures soar above seventy-eight degrees Fahrenheit, zebra finch parents are able to influence the speed at which their babies develop. When chicks hear this particular song, they slow down their development and are smaller when they hatch. Being prepared like this by their parents even helps them into adulthood, leading them to choose nest areas that are warmer, which is crucial, as many species have to adapt to changing climates as a result of human activity.[9]

At the same time, we can't prepare our offspring for every eventuality. Some things they have to learn on their own through trial and error. When it comes to knowing what to be afraid of, young prairie dogs have a lot of experience in the "error" part. If you have ever been on a prairie dog town, you know it is quite a noisy place. There is a lot of chatter, possibly jump-yipping (or sun salutations, as I like to call them, because of how they stand on their hind legs and throw their upper body back with front legs raised skyward), and sounding off the alarm. Communication among all the individuals in the "neighborhood" is critical to survival for prairie dogs. After all, they are on the menu for just about every carnivore found in their area. I had the good fortune of working with Dr. Constantine Slobodchikoff, who has done extensive research on the language of prairie dogs. Adults have a finely tuned communication system for warning others of an impending threat. In response to detecting a predator, individuals call out, using a highly specific call encoding what is threatening them, where it is coming from, and how fast it is approaching.

I spent a number of years of my life sitting on a rock, under a tree, in a tree, or atop a vehicle, all in an effort to better understand these marvelous animals. Every June a new batch of tender, young, and inexperienced pups would emerge from their burrows. These little prairie dogs haven't learned what to be afraid of or mastered prairie dog language. In those first months, a few things become obvious. First, young prairie dogs sound the alarm at *everything*. Second, they haven't got the words down right and quite literally call wolf when it's a hawk. And finally, the adults hardly pay any attention at all to the calls of pups. The stakes are high for young prairie dogs, where a mistake can cost them their life, so by the time August rolls around, all the babies that managed to survive are up to speed on prairie dog language.

Other lessons can be learned more progressively with encouragement from the mom and/or dad. A fellow academic shared with me how he helped his son take the leap of riding his bike to school on his own. The first few times, he rode with his son, leading the way. Once the route was familiar to his son, the next step was to casually suggest they ride together, side by side, to school. After a few more times, he asked his son if he could take the lead. And finally, one morning, he pretended that he was a bit tired and suggested his son go on his own. By incrementally increasing the level of his son's independence, his son could become more and more familiar with the process and grow in confidence.

This approach is strikingly similar to what I observed a pair of mourning doves do as they taught their chicks to fly so they could successfully fledge. While I've already made clear my admiration for mourning doves, watching these parents raise their family only served to increase my affections for them. As I've mentioned, every year a pair of doves would come to nest near my flat. Initially, they set up their nest precariously perched in a giant pine tree outside my living room. Two years in a row, the storms came in late March and toppled the nest, eggs and all. Finally, in the third year, the pair commandeered a planter I had hung up against the wall on my deck. It was the perfect spot, protected in the rear, tucked away from the elements, and offering a 180-degree view of any danger that might be lurking just off the second-story balcony.

I tended to let them be, watching surreptitiously from inside as the couple incubated their eggs, kept their newly hatched chicks warm, alternated feeding them, and finally fledged them. But one summer it must have been a particularly good year and the pair was on their third brood. I decided it was time to "take back" my porch and I began to sit quietly outside with them as long as it didn't appear to disturb them or alter their behavior. For the most part, when I was outside, the chicks, growing quickly out of the awkward stage, would hunker down into the nest, perhaps hoping to go undetected.

Watching the parents encourage their chicks to start flying was eye-opening. Prior to flight training, the mom and dad would land on the nest to feed the chicks—usually two—their food. It was beautiful to watch those little yapping beaks rise up, begging for food. Once satiated the chicks would disappear, only to pop back up when one of their parents landed. At some point, the parents decided it was time to start flying lessons. The first thing I observed was that the chicks began stretching and flapping their underdeveloped wings. Then, maybe a week later, rather than landing on the planter, the parents landed on one of my outdoor chairs that had a thick cushion and was in close proximity to the nest. Once situated they cooed and gurgled until the chicks got to the edge of the planter, at which point they increased the frequency of their calls. Sure enough, seconds later, the chicks awkwardly flapped and sort of crash-landed on the cushion. Then the parents reversed their position. They flew to the nest and called for their chicks to come back. Another week or so went by with the takeoff and landing getting easier and easier for the babies. The parents' next step was to increase the difficulty slightly by positioning themselves on the banister rather than the chair. After this, progress was rapid and the parents increased the difficulty first by landing farther away from the nest on the patio banister and then on the pine tree just off the balcony. About three weeks after all this teaching, everyone was flying to and fro with ease until, finally, they were gone.

As I witnessed this I couldn't help but think that this was the optimal way to help kids achieve their own milestones. The difference is that there are a thousand small, medium, and downright enormous markers along the way for humans. Maybe not all of them require this approach, but how long it takes to master any particular skill will vary from child to child. By adopting a mourning dove strategy and respecting the autonomy of each individual, parents can give children the opportunity to increase their confidence bit by bit, learn to trust their ability to accomplish unfamiliar tasks, have a sense of adventure, and be resilient in the face of difficulties. I think by and large, just as mourning doves and many other animal parents do, human parents adopt this approach when it comes to their children's physical accomplishments, be it learning to walk, dance, ride a bike, or find their way to school.

In other areas of development, though, we treat our kids more like barnacle goose goslings that breed high up on cliffs in Greenland. Barnacle geese, wherever they breed, choose to build their nests precariously perched on cliffs to protect them from predators. This strategy comes with a hefty price tag. Barnacle goose parents don't bring food to their newly hatched goslings, which means that everyone in the family must quickly get down the cliff to the ground where the goslings can feed alongside their mom and dad. They must eat within three days of hatching, but unlike their parents, they cannot fly yet. Therefore, when they are just three days old, the fluffy baby geese must take the biggest leap of their lives and catapult themselves off the cliff to the rocky ground below. And we are not talking fifty feet. It's more like four hundred feet. What happens? A lot of them die, but not all of them—otherwise there would be no more barnacle geese. What makes the difference between living and dying? It's not clear, but gosling size and amount of feathers are possible factors—as is, it seems, a fair amount of luck. Some goslings are more reluctant than others, and the difference in how long it takes some goslings to take the leap is relevant to understanding the difference between children.

One way in which we throw kids off metaphorical cliffs is by sending them to day care or preschool. Some kids take to it from day one and have no problem whatsoever, so the barnacle gosling attitude works just fine. Other kids may need extra time and extra coaching. Knowing that, accepting that, and working to get them slowly accustomed to changing situations can go a long way toward avoiding problems. For instance, a former colleague of mine knew her daughter, three-year-old Madeline, was an anxious, timid child who didn't take well to sudden changes. Understanding this and respecting her daughter's needs, she and her husband implemented a plan to get Madeline ready for transitioning from day care to preschool. Unlike barnacle geese who give their goslings no preparation time, Madeline's parents began talking to her about preschool about one month before she was to begin. Every day, they would talk excitedly about all the aspects that would be similar to what she was used to, and also everything different: new kids, new activities, new place, new teacher. You get the picture. At first, Madeline was suspicious and resistant, but after about a week, she started imagining what it might be like. Then she had three more weeks to get excited about it. What happened? Madeline adjusted quickly to the new routine because she had been given the opportunity to envision the change and look forward to it.

WILD LESSONS

* Teaching our kids about danger is important to preparing them for future success, just as zebra finches do for their chicks. At the same time, we must balance this so we don't make them afraid of everything.
* Incrementally increasing the complexity of a lesson (for example, how to get to school on one's own) helps kids build confidence.
* Many parents are patient when it comes to teaching their kids physical skills gradually, but then throw their kids into new situations with little preparation. This may be risky, depending on the personality and temperament of a particular child.
* It is easy to underestimate how intimidating new situations can be for some children. The mourning dove tactic of gradually exposing them, can go a long way to getting them to accept change as a normal part of life.
* Building children's confidence early will help them take on challenges later in life.

Learning to Get Along

Another big hurdle for kids (and sometimes parents) is heading off to school. Whether it's preschool or kindergarten, eventually most children have to spend part of their day away from their parents *and* enter this complex social arena called society. You may think that if your child went to day care there is no reason to be scared about preschool, or if he or she went to preschool there is no reason to be nervous about kindergarten. Yet so many children do experience stress, anxiety, or downright fear about these transitions. What's going on here? Why would they suddenly be clingy and reluctant to go off to school? Yes, some of it may be about change, as was the case for Madeline, but for the majority of children, what many parents may not realize is that each school represents a completely new social environment.

You might think you socialized your child in mommy-and-baby groups, playdates, day care, preschool, and so on—and you did. The problem, and this is why I think a lot of kids still have trouble, is that in each scenario they were socialized with a fairly consistent set of other children. That means they came to know certain individuals, the rules among a particular set of kids were established, and order (or chaos) was known. Oftentimes, with each transition, there is a completely (or at least partially) new set of children, on top of a new place, a new teacher, a new time, and so forth. We may be the only species that has to repeatedly integrate into so many different and novel social groups. And this is frightening. It is risky entering into a new social group. If you are in preschool, you may be with most of the same kids for two or three years before beginning kindergarten. In that time frame the social hierarchy is learned and negotiated, only to abruptly change when you go to kindergarten. Suddenly there are new kids, a new teacher, a new school, and you don't know where you stand socially.

Dwarf mongooses also experience this type of instability when they leave one social group and move into another one. Like us, they go from their home group, where they were born, to another one full of strangers. These small carnivores are found throughout Africa, living in burrows, and their group may have as many as thirty individuals. When the young dwarf mongooses disperse, or leave home, it is a risky time, and when they find a new group, there are often disadvantages to being the newcomer—specifically being at the bottom of the pecking order. Similarly, when a child moves abruptly to a new school and is the only new kid in class, he or she will often be targeted for bullying.

How can we help our kids adjust to these new social environments? One approach is to begin teaching children what the social rules are and how to integrate successfully into any environment by engaging in some very specific behaviors. Sharing comes to mind. This may come up even before school, such as on the playground. Sharing may involve toys, but frequently it involves food. Food sharing is a fundamental form of cooperation and it manifests early on. Kids often naturally share, but they may not want to share their *favorite* foods. This is also true of chimpanzees, or at least one I know particularly well named Kenya. Her home is still the Center for Great Apes, where she is now a gorgeous adult, but when I first met her she was about eighteen months old, and the center was located temporarily on the grounds of Parrot Jungle in Miami. I would take Kenya for walks on the park grounds before it opened and we regularly shared lunch together. What that meant was I would sit in her enclosure with her while she ate her lunch. She would get an array of vegetables: carrots, cucumbers, sweet potatoes, etc. But her all-time favorites were red or yellow peppers. She offered me any number of foods that ranked low on her list of favorites, but there was only one way she proposed to share her peppers: after thoroughly covering them in saliva. I don't know how she knew I wouldn't take them, but if I suggested I wanted a yellow pepper, she would put it completely in her mouth and then offer it to me.

There is some evidence to suggest that when children as young as three collaborate, they are more inclined to share equally as opposed to when they experience a windfall, in which case they share but keep the majority for themselves. To examine this, researchers gave a pair of kids some problems to solve that involved determining who had control over more than half of the toys, and then allowing that child to decide whether or not to split them evenly with his or her partner. For instance, in one experiment the pairs of kids were faced with a problem of accessing toys by pulling on a rope that would release the toys from behind an enclosed area. In the collaborative game, both children had to pull, but an extra toy would fall out, so one kid received three toys as a reward and the other only got one. In the scenario where they did not work together to receive the toys, once they entered the room, three toys were on one side and one toy was on the other side. Remarkably, when the pair worked together, the kid who was fortunate to accidentally receive an extra toy automatically gave it to the other kid—without prompting. When they did not work together, the child who received the extra toy did not spontaneously share.[10] This is strikingly similar to what we see in experiments with capuchin monkeys, who are well known for their sense of fairness.

This implies that, although we are inclined to share, we are still rather selfish unless we work together. Teaching children to share just for the sake of it is unlikely to resonate with all children. However, teaching children to work together toward a common goal allows for a natural expression of fairness.

Another valuable social skill is empathy. As with sharing, some individuals are more empathetic than others, and this is reflected in differences observed even at a very young age. The ability to empathize, to understand the emotional state of another individual, is related not only to sharing, but also to modulating aggression, since one who can relate can imagine the distress of another and is more likely to demonstrate helping behavior. The cognitive aspect of empathy continues to develop into adulthood, but young children have the emotional capacity and responsiveness to be empathetic and, therefore, helpful. Children as young as one year old can recognize sadness in other people and are motivated to help others feel better.

Rats may seem like an unlikely candidate for comparison, but their empathetic response to fellow rats is well studied. What does rat empathy look like? One study in 2011 found that when rats observed a fellow rat trapped in a tube, they rapidly figured out how to open the door to release their comrade, even if there was no reward and regardless of whether they were permitted to have physical contact with the rat they saved.[11] In another experiment, researchers placed rats in a pool of water, which is distressing for them. Next to the pool was a dry compartment. The only way the immersed rats could get out was if a fellow rat opened a connecting door. And those fellow rats did. And most of the time they even turned away treats in order to help the other rat out.[12]

There is an unexpected twist in the rat empathy story: Rats do not discriminate when it comes to whom they help or don't help, unless their social environment is not diverse. What does that mean? First, rats develop biases depending on their social experiences. They will definitely help a rat with whom they have had a past positive experience. Second, rats' coats come in different color patterns. Some are white, some brown, some black, some spotted. And if rats are raised exclusively with rats that look just like them—white only, or spotted only, for example—they will refuse to help rats that look different from them![13] The implications of this are far-reaching because we know that humans also show similar biases. I am quite certain that we're seeing similar mechanisms at play in both rats and people. Exposing children to a variety of people, places, and cultures reduces the tendency to see any one group as different and broadens the scope of empathy to include all people. This means that even though our tendency to be helpful and empathetic is part of our biological inheritance, it needs to be nurtured, for some people more than others. Who is responsible for that? Parents.

If you have a child who seems to be less than naturally empathetic, what should you do? One thing to realize is that children who feel secure emotionally at home and receive all the love and support they need are more likely to give to others.[14] Children also develop empathy, and other skills, by modeling their parents' behavior. Expose children to a variety of people and situations, because they are more likely to feel empathetic toward something that is familiar. And lastly, experience is a big component in the development of empathy. Children who are exposed to conversations or even books about the perspectives, feelings, and experiences of others strengthen their ability to empathize.

Social skills—especially empathy—become incredibly important once children enter school. Research suggests that a lack of empathy goes hand in hand with bullying. Bullying is inevitable if parents don't take responsibility for training their children in empathy. Bullying is about social or psychological dominance, and it's common in despotic or tyrannical societies.

Baboons show the terrible suffering bullying inflicts. Studies have shown that, among baboons, bullied individuals experience increased stress and lower immune response, and they develop neurotic behavior patterns that can ultimately lead to death. As for bullied humans, the fact that some resort to suicide is evidence enough of the potentially serious effects of bullying.

Bullying is a part of life for banded mongooses, whose social structure is characterized by a strong dominance hierarchy; dominant females will harass and even kill the offspring of subordinate females. Social status in other species often comes with big benefits, such as increased access to resources—be it food, friends, or mates. So there can be a payoff for bullying.

The question is, do we humans also benefit from bullying one another? The answer would seem to be yes. The rate of bullying among humans is fairly high, perpetrated by an estimated one hundred to six hundred million adolescents globally and from all cultures every year.[15] And that is just adolescents.

If we threw adults into the mix, those numbers would be even higher for a behavior that we *think* is maladaptive, meaning we don't believe it provides enough payoff to make it worthwhile. We even spend millions in campaigns to eliminate bullying. Bullying isn't new in humans, and there is some evidence that in ancestral and contemporary populations, bullying gives individuals an edge in much the same way as it does for the banded mongoose: increased mating opportunities.[16] Therefore, some might argue that bullying is a natural consequence of being human. On the other hand, there are plenty of cultures, specifically egalitarian hunter-gatherer cultures, that are characterized by a way of life and a way of thinking that honors the autonomy of each individual, where no one is coerced into any behavior, individuals are respected for their contributions, and judging oneself as better or worse than another is not supported by the community. What does all of this mean? It means that we, as humans, can create the social and environmental conditions that either foster empathy or support bullying. As parents, we directly influence the trajectory of future human society and culture.

WILD LESSONS

* To some degree, sharing comes naturally to kids. Parents can promote equal sharing by having children work together.
* Empathy is a skill that is nurtured all through development. Parents can promote deeper empathy by talking about the psychological state of others and exposing their children to a range of people, places, and experiences.
* A lack of empathy and bullying go hand in hand. If your child is a bully, then take steps to train him or her in empathy.

Time to Fly

There comes a time in every family's life when it is time for the child to move out! Sometimes kids want to be independent before they are ready, and sometimes parents want their kids out before the kids feel ready. As with weaning, going to school, and gradually developing one's social chops, the transition to independence can be a rocky one indeed. Believe it or not, examining the conditions under which offspring leave home or stay is a major question in the study of animal behavior and ecology. What factors influence these decisions and when, and even whether these decisions are made at all, is quite variable across species.

For some kids, their mom or dad gives them the old heave-ho whether the kids want to leave or not. This is certainly true for some humans and for mute swans. After the cygnets have undergone the plumage change and their downy, gray feathers are replaced with their gleaming, white adult feathers, their dad begins chasing them. He relentlessly harasses his kids, forcing them to leave and strike out on their own. Why does he do this? In the case of mute swans, before the parents can mate again, build another nest, produce a new batch of eggs to incubate, and raise another family, the previous generation has to go. Human parents may not use the reasoning that they want *more* kids, but some have a definitive idea of when to pull the plug on parental support in terms of providing a place to live or resources such as food or money. In the United States, when a child reaches the age of eighteen, he or she is considered an adult. At that point, some parents choose to evict their offspring and force them out into the world on their own, swan style.

But are eighteen-year-old human children truly equipped to make it on their own in the society we have created? Some may be, but others still haven't mastered the first skill discussed in this chapter: feeding themselves! Okay, yes, they know how to use a fork, but they don't necessarily know how to cook, make a budget, or tackle any number of complex adult tasks. Recently I was grabbing an afternoon coffee on the college campus where I work and struck up a conversation with one of the students behind the counter. I had observed a large group of seventeen- to eighteen-year-olds wandering the sidewalks with people who were clearly their parents. I asked him what was going on. He informed me that it was a visiting day for prospective students and their parents. And then he said, "I remember when I first came here for school. It was the scariest time in my life!" I replied, "Really? Was it?" "Oh yes!" he continued. "I had never been far from home and I wasn't even sure I could leave home at all." I was curious just how far from home he was, so I asked. His response: a two hours' drive.

Not all kids are ready to strike out on their own at such a young age, and, in keeping with the theme throughout this book, this translates into continued parental investment past the arbitrarily defined legal age. Yet historically (and in many societies still today) older siblings were expected to begin helping and working toward the greater good of the family and community by age six. By the time they reached puberty, the question of staying or leaving may have depended heavily on how much the now sexually mature offspring contributed to the group compared to how much they consumed.[17] This consideration is encapsulated in the common line parents use today when addressing their legally adult children: *If you stay under this roof, pay rent and contribute to the household.* When there are younger siblings still in the house, this can even mean helping to parent in some capacity.

This dynamic is mirrored in superb fairy-wrens, common to Australia and typically found in lush wooded areas, where older siblings stick around, and in exchange for getting the benefits of remaining in their natal, or birth, territory, they help their parents out by assisting with raising the next generation. A big challenge young fairy-wrens face when they are grown up is finding a place to live,[18] not unlike many adolescent humans, especially in places where housing is limited, in which case, the age a young person can leave the house can extend into the mid-twenties or older.

Some parents don't push their kids out at all, nor do they demand any contributions in order to stay. Roe deer, common to Europe, need a bit of extra time before striking out on their own, and mothers let their older offspring hang around until they are ready to disperse. What determines whether or not a young deer is prepared to strike out on his or her own? Body condition and social aptitude. How big the deer is and the likelihood of being able to emigrate into a new social group and survive the winter are the determining factors.[19] The difference between roe deer and humans is that although both experience some competition for access to food, a roe deer mother does not provide food, nor any direct parental care, for her older kids.

An interesting difference between us and most other species is that once other animals disperse, they rarely come back home. We, on the other hand, seem to "fly back home to the nest" quite often. This is not a new phenomenon, and perhaps our rapid departure from the nest is only suitable under just the right economic conditions. The data suggests that when times are good, we fly out of the nest much earlier than when times are tough. Not only do we depart sooner, but we disperse farther away from home.[20] The rate of adult children returning to live with their parents began increasing in the 1980s, and by 2012 roughly 10 percent of households in the United States had adult children, aged eighteen or older, present.[21] Whether or not this trend will continue may depend largely on economic conditions.

It may happen more frequently than we realize in other species— this coming back home because things didn't quite pan out—but the data is sparse. Among humans, not all parents welcome their children back with open arms. And if they do allow an adult child to come back in, there can be enormous conflicts over independence, rules, threats of eviction, and how much the adult child needs to contribute—your standard superb fairy-wren arrangement. The best option for offspring is to take advantage of parental investment as long as necessary, but not to overstep a parent's need to make the child fly!

CHAPTER 8

No!

The Nature of Discipline

My friend Paul is, as I mentioned, a thoughtful person. He's also a deep thinker. With three children at home and a research focus—the evolution of social behavior—similar to mine, he has co-opted his family as a test bed for the concepts of cooperation, conflict resolution, and parenting. He recently remarked, "If I have one word to describe parenting, it's 'boundaries.' Setting them, enforcing them, and maintaining them." To me, the matter of boundaries falls within the realm of discipline. But what, exactly, is discipline? Is it simply keeping kids within particular boundaries, or is it about teaching them new skills? Is it for their own safety, or to serve the needs of the parents?

These questions matter because if, as a parent, the goal is to teach a child, then educational approaches would make the most sense. On the other hand, if the goal is to command obedience through dominance, then aggressive physical punishment is an excellent strategy. And while I wouldn't recommend the latter, in other species we see punishment emerge as a strategy to enforce social norms and dominance structures through fear, intimidation, retaliation, and spitefulness. The question, then, is: Are parents tyrants or teachers?

Punishment and policing do happen in otherwise cooperative species, and it has been suggested that this is necessary to encourage cooperation by reducing the number of cheaters, or those who fail to follow the rules. Case in point: the emerald coral goby. Few may realize that, like their counterparts the clown fish, coral reef gobies have complex societies. Emerald coral gobies lives among the anemones and corals in the Indo-Australian Pacific archipelago, where they are protected from predators and, in turn, assist their friendly hosts with reducing parasites. Consider a single coral a house in which you might find a pair of gobies that are breeding and a few individuals that aren't. Whether an individual breeds or not is tied to body size, and this is especially the case in fish. Now, throw into the mix the fact that "houses" might be few and far between (with lots of hungry predators lurking) and you have a recipe for cooperation and an it-pays-to-stay attitude. But this comes at a price. If you can't reproduce, you have to stay small. If you break this "don't reproduce" rule, the punishment— enforced by the dominant breeding pair—is eviction![1]

What is noteworthy about this example is the social environment where the policing occurs. Punishment is directed toward other *adults*, not offspring. Meting out punishment comes at a cost. Essentially, the punisher pays a price for retaliating (lost energy, risk of injury) against a noncooperative individual, and this cost is only offset if the cheater changes his or her ways. What this suggests is that punishment is only worth the cost if it actually works. Ultimately then, punishment is self-serving and selfish, which may explain why parents who choose to spank their children might utter the words, "This hurts me more than it does you." It doesn't *really* hurt the parent more, but we'll investigate this further below.

So what does this tell us about the nature of discipline? Are animal parents punishing, teaching, or both? Does the form of discipline vary with age? Some human parents seem to have behavioral expectations of their children that far exceed their mental, emotional, and physical abilities. Do other animal parents have difficulty assessing what their offspring are capable of doing? We'll explore these questions in this chapter.

Many people feel that without punishment children would run amok, wreaking havoc on everyone's lives, or worse, that children would fail to learn the social rules (essential to cooperation) and face punishment later in life. For example, some parents may believe that by punishing their children severely they are ultimately saving them from being punished by others in society, including the police.

There is little empirical evidence to support the idea that aggression results in learning—and that goes for humans as well as other animals. Instead, aggression is more likely to cause the individual on the receiving end to mitigate the stress and fear that punishment induces by either obeying or avoiding the aggressor. Basically, if children comply it's because they are *afraid*, not because they are learning effectively. But are there situations where we are less concerned about learning per se than compliance? And if we want our children to comply with our rules, is the threat of severe punishment the most effective means of achieving this compliance?

There are many disciplinary approaches for parents to navigate. Check the parenting advice section at your local bookstore and you are guaranteed to find different perspectives on the pros and cons of various approaches, along with confident directives for parents: *Allow your child to explore relatively unrestricted! Help your child develop the self-discipline needed to succeed! Do both!*

Which way is best? Should you pick a strategy and stick to it? After asking yourself whether you are trying to teach or dominate, a more pertinent question might just be: Do you know why you are saying yes or no to certain things? Barring situations of imminent danger, are you so busy saying no that you are missing opportunities to teach your child important lessons? How do animal parents know when to say yes, when to say no, and when to ignore the antics of their offspring?

Are there animal "helicopter" parents? And why do some parents, animal and human, take discipline too far and harm or even kill their offspring?

How Old Are You?

It was a sunny morning. The air was crisp and clean, a gentle breeze blowing across my face as I sat outside my favorite French bakery waiting for my breakfast. I had forgotten it was Mother's Day and the bakery was teeming with families. Mothers and fathers, with their children in tow, filled the place. A tent stood outside the entrance where servers handed out a special gift just for moms. I imagined that inside the little box there was a melt-in-your-mouth, delectable pastry, warm and sweet.

My preoccupation with the contents of the box was interrupted when a family sat down at the table next to mine. I don't recall what the mother or father looked like, but I won't forget the little boy. He couldn't have been more than two years old, with red hair and a face peppered with freckles. They put him in a booster and he sat next to his mother at the tiny bistro table.

That day mothers also received a special drink, a mimosa served in a tall, slender champagne glass, to complement the meal. My brunch arrived, and as I dived into my omelet, I noticed that things were getting pretty crowded at this family's little table—coffee, water, a special box, a special drink . . . perhaps you are already anticipating where this is headed. The toddler was restless. He flailed his arms around and made contact with mommy's special drink.

What happened next was completely unexpected. Both parents yelled at him, telling him what an awful child he was and how he had ruined his mother's special day by knocking over her special drink. The confusion that came over his face was swift, followed immediately by anguish and tears, undoubtedly in response to their anger. My heart broke for this little freckle-faced boy who was uncertain about what exactly he had done wrong, but was clearly devastated by his parents' reaction to him. My only thought was, *If I could have seen that coming, why didn't they? And how on earth can you expect a two-year-old not to knock over stuff?*

Perhaps to their adult brains this made perfect sense, that he could make the connection between accidentally knocking over his mommy's drink and being screamed at, and discern what he was supposed to learn. I suppose one could argue that their confidence in his cognitive abilities was admirable, but I suspect there were other possible explanations. Perhaps these parents didn't want to accept the trade-offs associated with parenting. For example, it might not be possible to have a sit-down brunch in a crowded, cramped café with their two-year-old without having these sorts of challenges. Or maybe these parents were ill-informed about child development. From a purely scientific perspective, I could see that in this situation not an ounce of learning would occur—other than learning to be frightened. Frankly, the concepts of a "special day," of "ruining" something, and of "proper behavior" are lost on a two-year-old.

If we set aside whether their motivation was a consequence of ignorance, frustration, or a refusal to accept the constraints that come with taking young children into certain environments, and whether they intended to exert dominance over their toddler through fear and intimidation or to teach him to be more careful, this incident provides an excellent springboard for discussing age-appropriate discipline.

For animals, understanding this concept is incredibly important. Just ask a chimpanzee. You may recall from Chapter 3 that young chimpanzees have a white tuft of hair adorning their backsides. This marker is a signal to everyone in the community that this is a juvenile who is still learning the ropes about what it means to be a chimpanzee. Although all members of the community display a high degree of tolerance in response to the antics and intrusions of immature youngsters, chimpanzee mothers are especially indulgent of the shenanigans exhibited by their young offspring. One situation ripe for a youngster's interference is when his or her mother is fishing for termites. Termites are an important food source for chimpanzees, but getting their hands on them can be tricky. To accomplish this, they use a stick or twig that they have to modify in a particular way, insert their tool into the termite mound and, once some termites have attacked the "intruder" by clinging onto the tool, carefully extract the twig and swiftly munch the tasty termites.

Take a moment to think about something you were doing recently that required just the right tools and intense concentration. Got it? Okay, good. Now, imagine your two- to five-year-old peering inches away from your hands or face, grabbing at your tools, or worse, pulling on your hair, trying to nurse from you, demanding your attention, or otherwise disturbing you. How do you feel? Aggravated? Annoyed?

This is the life of a termite-fishing chimpanzee mom. It takes young chimpanzees quite a bit of time to get the hang of all the steps needed to successfully catch their own termites, with juveniles not achieving success until they are two and a half to five years old.[2] Aside from passively watching, young chimpanzees also do everything from reaching for their mom's hand, mouth, or tool to stealing the tool outright, to sniffing, peering, or poking around and in the mound, to trying to grab the termites off her tool. That is, if they are even focused on learning the task. Chimpanzees this age also get distracted easily, may become bored, or insist on playing during "work time."

When researchers examined how mothers who were termite-fishing responded to being bothered by their offspring, they found that the vast majority of moms respond by *ignoring* their offspring's intrusiveness—a whopping 85 to 86 percent of the time. When a mother does have a negative reaction, she moves away from her kid, gently pushes the kid away, or changes body position (e.g., by turning her back). It seems that, for the most part, chimpanzee moms are adept at "tuning out" their young without treating them as they would older offspring or other adults. And so they end up being incredibly tolerant of their children's behavior. That being said, there is still an upper limit to how annoying a youngster is allowed to be before some moms deliver an assertive *enough!*—which, in this case, is a gentle push away from the mound. Boundaries.

One of the primary functions of this kind of tolerance is to facilitate and support abundant opportunities for learning. Because this leniency is linked to age—chimpanzee moms become less tolerant as their children grow—we can infer that other animals likely can differentiate stages of development and ability and treat their offspring accordingly.

Like chimpanzee moms, many human mothers and fathers are extremely patient when it comes to teaching their kids how to play catch, hit a ball, ride a bike, tie their shoes, use the toilet, etc. But what about a situation where learning a skill is not the objective, where your child simply wants your attention or something that has caught his or her eye? Let's think about the supermarket aisle. Recently, I was in Colorado to film prairie dogs for research, and I made a run to the local Target for some snacks. There, I came upon a scene played out all across America and beyond: the tantrum-throwing toddler who did not get the candy, cereal, or toy she wanted. This little girl was maybe three years old, a ripe age for pitching a fit, with just the right amount of shriek in her voice to complement her rage.

Her father, in response to her collapse to the floor, initially began a complicated conversation with her, explaining all the rational reasons why he would not buy her the item that caught her attention. Her response? Predictably, she became more hysterical, shouted that he was the meanest daddy ever, and writhed in agony on the floor.

Sadly, as many tantrums do, this one escalated into World War III. Dad lost it, threatened her with anything he imagined would distress her, spanked her (causing an even greater meltdown), and finally grabbed her and screamed within an inch of her face. The scene ended with him picking her up and storming out of the store as she shrieked, squirmed, and cried.

This kind of incident normally happens with children aged two through four. Naturally, there is some variation in temperament among children—not all experience meltdowns. However, in general, the ability to control impulses and self-regulate emotion is not something that children this age are actually capable of doing. Let's look at why a bit more closely.

The primary area of the brain responsible for the ability to regulate what happens when a desire for a sparkly toy or sugary treat is denied is the prefrontal cortex (PFC). This part of the brain is one of the last areas to develop in humans and only begins maturing around four years old. When does it stop growing? In the mid-twenties. (This fact is relevant not only to the discussion here, regarding toddlers, but also to risk-taking behaviors in teenagers.)

This region of the brain is rather complex, with connections to all sensory systems as well as brain areas involved with emotion, memory, and motor function.[3] It also makes up about one third of the neocortex, and so it's heavily involved in cognition, self-control, and language acquisition, too. In other words, it has its fingers in a lot of places. How does this relate to the apple of your eye losing his or her mind? Unfortunately for your little one, he or she is besieged by all kinds of sensory information (sights, sounds, smells, tastes, and tactile sensations) without the power to control any automatic responses he or she may have to these inputs, be they desirable or disagreeable.

Imagine, for a moment, that your world—that is, your inner world—is receiving all the stimuli of the world around you, and that you're not able to control your reaction at all. Then pile on top of that primitive language skills and a lack of comprehension about complex societal norms. A toddler can completely comprehend the surface layer of most things. For instance, a toddler will have no trouble grasping the concept of having ice cream. That he or she can't have ice cream until after dinner is a bit murky. And that ice cream is not to be eaten in lieu of dinner in order to establish healthy eating habits is completely lost on the child. These abstract, deeper explanations are meaningless to toddlers. Expecting otherwise is unrealistic.

As challenging as it might be to cope with a toddler-turned-Tasmanian devil, imagine what life must look like in his or her head. Until approximately the age of seven, children have difficulty differentiating their worldview from anyone outside of themselves—and even then, their ability to do so diminishes as the number of distractions increases.[4] This is not a license to let them persist in this line of thinking or behavior, but a balance must be struck between where kids are developmentally and where parents want them headed behaviorally. What does this mean for parents? In part, it means recognizing there are biological limitations to what toddlers and young children can comprehend and what they can do to regulate their own behavior.

At the same time, it would be remiss of me not to acknowledge that toddlers naturally test boundaries and try to determine their own limits. One way they do this is by throwing tantrums. Other species' young have tantrums, by and large, when weaning. Like in humans, these outbursts may be used as a form of manipulation. Toddlers are aware that they are smaller and weaker than their parents. A good option, then, as they see it, is to pitch a fit.

The same is true for stump-tailed macaque (or bear macaque) youngsters, who throw tantrums when their moms are weaning them. Like other macaque primate species, the stump-tailed, or bear, macaque, lives in fairly large mixed-sex social groups with a strict female hierarchy. They look like miniature Sasquatches (if Bigfoot actually existed) with reddish faces. Females enforce hierarchies with some slaps and bites, while males tend to take a more "let's all get along" approach. (As an aside, if males do get into tussles, the subordinate individual will apologize by presenting his rear end. The dominant individual will acknowledge this gesture with some teeth-chattering and lip-smacking. I wonder if that would work for us? But back to tantrums.)

Just as with human toddlers, young stump-tailed macaque tantrums can be motivated by *I want what I want, and I want it NOW!* And what they want is continued access to their mothers' milk. This may be a genuine need or, as many parents can relate, the beginnings of testing the limits. When denied, stump-tailed macaques whistle in protest. It's not clear if these tantrums work for stump-tails, or if their moms basically ignore them and go about their business. Research suggests that stump-tailed macaques are somewhat lenient moms, possibly owing to the fact that other females participate in taking care of and paying attention to juveniles that are not their own. Perhaps this extra attention from other females helps mothers avoid being overly concerned with their offspring's tantrums. At the same time, perhaps they are more relaxed about establishing parental boundaries, something all offspring, human or animal, have a strong instinct to explore.

When it comes to being tolerant, the same cannot be said for rhesus macaque mothers. How many parents out there have given in to their toddler's rant because they were in public and others were witness to their behavior? Rhesus moms, for one. We tend to think about the "bystander effect" as one's apathy or unwillingness to act when witnessing someone being victimized because other people are present and one presumes that someone else will take action. In the case of tantrum-throwing rhesus macaque juveniles, the bystander effect refers to how mothers react differently to their own offspring depending on who is around. Female rhesus macaque dominance hierarchies are characterized by tons of aggression, and unlike stump-tailed macaques, female rhesus macaques may kidnap and harm a subordinate female's offspring. One question researchers asked was: Would a mother give in more often to her infant's weaning tantrums if she were in the presence of others who might be potentially aggressive toward her because of her crying infant? And no, this experiment did not involve putting rhesus macaques on an airplane!

But first, what does a baby rhesus macaque tantrum look like? Awfully cute and funny to me—but then again, I am not its mother or a rhesus macaque. First, the baby cries and whines. If this does not work, the tantrum progresses to a full-on meltdown, arms flailing and cries escalating to screams. Sound familiar?

Researchers discovered that mother rhesus macaques were more likely to cave and give in to their infants' demands when dominant, potentially aggressive individuals were around.[5] More specifically, if a mom was alone or around family, she acquiesced roughly half the time. But when someone higher-ranking was around, a mother gave in a whopping 80 percent of the time.

Why? Because when dominant, aggressive, unrelated macaques were around, they would shove, bite, and kick the infant and his or her *mother* in response to these tantrums. For the infant, the payoff in throwing a tantrum to secure milk could be short-lived because even the mom was more aggressive to him or her when there was a chance that a bystander would retaliate against her. This was fairly uncommon, but it does illustrate a subtle, salient point about the use of *physical* punishment against one's own offspring. Such punishment is very rare in other species and extremely context-specific. In this case, the mother rhesus macaque may be aggressive toward her own offspring because a much larger threat from another aggressive adult exists. It is the equivalent of yanking your child out of the street and lifting him or her up roughly out of fear that he or she will be injured. And yet, even though for the young rhesus macaque there is a real and pressing danger of attack, it is still *rare* that the mother will be physically aggressive toward her own infant. The purpose is not to teach—but rather to protect.

In humans, bystanders also get annoyed or aggravated, and many a parent has had to deal with the dirty looks, snide remarks, or other derisive comments of other adults. Two interesting elements come up here. First, why do bystanders, macaque and human alike, become perturbed by the screaming of someone else's child? Second, why do so many human parents react by becoming verbally or physically aggressive with their children in this context? For the first part of the question, it is a bit of a mystery why anyone would care. Setting aside the grating sound of a child shrieking at a frequency designed to grab one's attention, I propose that how a parent responds to his or her child may be driven more by the social environment of adults than the behavior of the child *per se*.

Stick with me here. When we took a look at the societal structure of rhesus macaques, we saw that it is only when a dominant individual is around that the mother caves as a way of extinguishing the tantrum. How does that relate to us? We humans have a tendency to regulate how parents parent. We expect parents to regulate their children in a particular manner, to effectively enforce the boundaries of what we collectively have deemed acceptable social behavior. When we observe parents failing to do so—failing to keep their kids "in bounds"—our response is to police the parents' behavior. We may not bite, kick, or slap, but we glare, sneer, judge, and make contemptuous statements, usually about the parents and their competence. In this sense, we are not so different from the dominant macaque, whose ultimate concern is ensuring that the offspring of subordinate adults realize that they too must "act subordinate."

This may also explain why so many parents attack their children verbally or physically under such scenarios. Although this is not the common response by rhesus macaque mothers, who generally just give in, we are infinitely more complex in our societal expectations and cultural norms of behavior. We humans far surpass other species in our desire and willingness to impose and enforce social norms, even ones we aren't certain reflect our own beliefs, and even in the face of behavior that does not directly harm anyone. Combine this with the importance we place on being viewed positively by others and meeting societal pressures for success, and it's no wonder that parents are quick to eliminate an "offending" behavior so as to avoid the scorn of other adults rather than deal effectively with their child.

There is solid evidence that harshly punishing children verbally (humiliating, insulting, threatening abandonment) and physically (slapping, kicking, dragging), especially in the early years, changes the course of cognitive and verbal development—for the worse. Not only does it reduce school performance, working memory, and higher-level cognitive functioning later in life, but it also delays language acquisition in small children.[6] Why language? It all comes back to the developing prefrontal cortex, which is extremely involved in language development, especially during childhood—one of the reasons why young children can learn multiple languages much more easily than adults. So any damage to it early in life is particularly consequential.

By now you may be thinking, *Well, this is all very interesting, but then how do I deal with my kids when they are having an outburst?* What we can see is that, for humans and other species, developing impulse control doesn't happen right away and overlaps significantly with the natural and necessary drive to test boundaries. The animal behavior research on self-regulation of behavior, restraint, patience—whatever you want to call it—has all been conducted on *adults*, not juveniles. Why? Because it is widely understood that juveniles do not have the same capacity as adults. This ability develops incrementally over a long period of time.

Consider dogs. Have you noticed how puppies show limited self-control? Comparatively speaking, one study showed that of the five adult dogs tested, all five were able to defer one reward for something they wanted more, and some of the individuals were able to delay gratification for up to ten minutes.[7] You may be asking yourself, *How did the puppies fare?* Puppies weren't tested because, well, they're puppies, and most of us—scientists included—know that puppies have little to no restraint.

But puppies and toddlers aren't nearly as bad as pigeons. I know some people don't respect pigeons, but they are pretty remarkable birds. They navigate using the Earth's magnetic field, they saved the lives of thousands of soldiers during World War I and World War II by carrying messages across enemy lines, and they mate for life. What they are not great at is impulse control. They are pretty horrible at it, really. When eight white king pigeons were tested for their ability to delay gratification for fifteen to twenty seconds in order to receive a greater reward, half of them had to be excluded for a variety of reasons, including not being able to focus on the task at hand. The remaining pigeons were only successful 6.6 percent of the time in waiting for the better reward, and only if they could *see* the reward.[8]

Human toddlers are able to delay gratification, but they need assistance from their parents. In one test, when a group of two- to three-year-olds received encouragement and explanation from their parents about the benefits of postponing opening an ordinary gift to wait for a better one, 75 percent of the toddlers chose to wait. Interestingly, 66 percent of toddlers whose parents simply actively demanded that they wait, didn't. It seems that when toddlers understand what is expected of them and why it is important, and are supported positively in making choices that align with desired behavior, they act as desired . . . most of the time.[9]

Thus, we can see that children must be taught, gradually and repeatedly, how to delay gratification, regulate their behavior and emotions, and reason properly. How does one accomplish this? It depends on the situation and why any given child is acting out. Is the child having an impulse challenge or pushing the envelope to see what he or she can get away with? Does it make sense to have the patience of a chimpanzee and ignore your child's tantrum?

Some experts recommend exactly that.[10] Assuming everyone's safety is secure, and the cause of the tantrum is attention-seeking or object-oriented (toy, sweets, etc.), your best bet is to ignore it. Seriously. A tantrum is almost always self-limiting. Ignoring it is like withholding oxygen from a fire—it will extinguish itself. You cannot reason with very young children because they do not have the *capacity* to reason fully and definitely not in that state. Fortunately, *you* can. Thus, the optimal strategy is to let your children know that when they compose themselves they will regain your attention, and then . . . wait it out. I know that three to four minutes of screaming and crying seems like an eternity, but it's not. That is the average duration of a tantrum, depending on how a parent responds and has responded previously. The limiting factor for a child isn't exhaustion; it's hitting a wall of no engagement. This strategy applies even when children are testing limits. Why? Because even kids have limits on how far they are willing to go to test boundaries.

Now, unlike most other species, we may have other children of different ages in tow at the same time. The result? Potential chaos. Therefore, it is the parent of the single child that may enjoy the luxury of ignoring a tantrum without disrupting the event, activity, or lives of the other children. This was pointed out to me by a single mother of three—two boys and one girl. She indicated that the youngest child learned to initiate tantrums during events and activities of the older children, which made it difficult for her to deal with just that child's behavior exclusively. This was particularly true if she was at a school play, a sports game, or in the car, running late, or dropping older ones off at school. Therefore, her usual strategy of ignoring tantrums was rendered ineffective, according to her.

What other options are available? Applying developmentally appropriate expectations to each child individually is a start. Perhaps a three-hour school play is beyond the abilities of a young child to sit through. Maybe like the young chimpanzees who lack the patience and focus to let their parents catch termites effectively, young children get bored watching their brother or sister play soccer. Coordinating activities with multiple children at different ages is a challenge, but I still think dealing with these outbursts is no different. One could even recruit the older siblings into modeling and encouraging appropriate behavior or ignoring inappropriate behavior. Once again, I think parents are more reluctant to let tantrums run their course when they fear the judgment of onlookers.

Perhaps, then, it would also help if all the observers out there stopped judging parents when their young children are having meltdowns. Yes, the whines and cries and unintelligible words are incredibly annoying, piercing the concentration of parents and bystanders alike,[11] but these aren't the result of bad parents raising ill-mannered, spoiled brats. These situations that parents find themselves in, often quite publicly, are a necessary part of raising children. What bystanders are frequently witnessing is not poor parenting, but rather boundary enforcement—which is often instigated by a child testing those boundaries. It is normal. Judging a parent for the fact that their child is having a tantrum is misguided.

Things change, though, when the reason for the tantrum is that your sweet innocent one is simply defying your wishes. Pandas tend to get a bad rap as parents because, both in zoos and in the wild, they accidentally sit on and squash their tiny, jelly-bean-shaped babies. But, once their baby has begun to resemble an actual panda and is scampering around in cute, furry panda fashion, it's a different story. I recently saw a video of a captive panda mom at a facility in China trying to get her young cub to turn in for the night. This feisty, full-of-energy bundle of adorableness was having none of it. After a few attempts to encourage her cub to come to bed, she had no choice but to resort to scooting him, pushing him, and standing over him while guiding him in forward motion, and finally picking him up and holding him. She did this gently but forcibly, balancing his weakness with her obvious strength. She showed restraint and remained in complete control of herself and got him to do what she wanted by doing it together.

Interestingly enough, this is also the approach recommended for human toddlers who are refusing to do something that they must do. Let's say it is time to get dressed and your child doesn't want to. You give your child a few seconds to comply, warn your child that if he or she does not, you will then physically assist him or her in getting dressed. This help should not be physically rough: being rough serves no purpose when you are fully capable of firmly but gently putting on your child's clothing. The important thing is that your child understands that getting dressed is mandatory, not optional.

To effectively discipline children there are three necessary ingredients: a strong positive learning environment, preemptive teaching designed to increase desirable behaviors, and corrective and clear reprimands (time-outs, verbal reprimands, or the taking away of privileges).[12] However, regardless of the child's age, time-outs only work when they are about a clear enforcement of boundaries.

The goal of parenting is to raise happy, healthy, and successful children who go on to be productive adults. Discipline is about teaching them not only how to regulate their own behavior, but also how to be a functioning, cooperative member of society. By supporting them positively, clearly, firmly, and with reasonable age-appropriate expectations along the way, you are empowering them to do just that.

WILD LESSONS

* As with other animals with a prefrontal cortex, the development of this "impulse control" part of the brain is the slowest to mature in human children.
* Children under the age of five have tantrums because they can't adequately deal with conflicting impulses and social expectations.
* Macaques and chimpanzees ignore the vast majority of their kids' meltdowns. Try it.
* Physical aggression toward your offspring doesn't stop tantrums or teach them how to behave. Instead, it *slows* their cognitive and language development.
* Using humor is one of the most effective strategies to foster learning in youngsters. Laughter reduces stress (for you and your child) and promotes connection and closeness.
* When physical aggression occurs in other species, such as rhesus macaques, it is extremely rare, context specific, and serves to *protect* their offspring from harm.
* If you observe a parent struggling with a tantrum-throwing child, don't judge. Instead, realize that the parent is setting and reinforcing boundaries and that the process isn't always pretty.

Come Here! Come Here! Come Here!

A lack of impulse control does not only manifest as a tantrum. One of the most ubiquitous problems faced by animal parents—including humans—is how to keep their offspring close and out of harm's way. Last summer I was observing a pair of Canada geese whose first batch of chicks had just hatched. There were six feathery, fluffy bundles of wobbly baby geese dutifully following their parents around. Soon after, I saw both adults in the street, frantic. That is the only word I can use to describe their behavior. They were flapping their wings, running back and forth in circles, and vocalizing loudly. There were no chicks to be seen. None. I imagined that each parent was squawking at the other, "Where are the kids?!" I never saw the chicks again.

Just as with young animals, once young children become independently mobile, the *come here* or *don't go there* struggle begins. When I was living in Arizona studying prairie dogs, there was a story in the news about a young girl, about three years old, who had wandered away from her parents at their campground. Unfortunately for this little girl, the end result of getting lost was being killed by a mountain lion. This is an extreme example of the worst that can happen when a child gets lost, but this was exactly the sort of danger our ancestors probably faced for millennia[13]. But even though, by and large, we are no longer surrounded by predators waiting to snatch up our children, there are still a lot of dangers out there. For example, cars in the street, holes in the ground, and kidnappers on the prowl all represent potential risks to offspring in the contemporary human environment.

Children do get lost and can be abducted. However, the vast majority of children reported missing end up being quickly reunited with their parents, or have been abducted by a relative or someone else they know. Only about 3 percent of the 900,000 children reported missing in the last two years went missing because they either got lost or injured or were abducted by a stranger. That comes to 27,000 children in the past two years.[14] There are roughly four million new infants born in the United States every year. All in all, then, we are doing a darn good job of keeping our kids close.

Nevertheless, young children are curious, fast, full of energy, and like to play. Hide-and-seek is a favorite game among toddlers. Perhaps they love the thrill of feeling momentarily independent, testing the boundaries of autonomy, while fully confident that they will be found and reunited with their parents or siblings. The danger is that young children don't know when it's safe to hide from their mom or dad or which hiding places are safe. They don't have the developmental ability to understand both context (when some games are safe and others are not) and consequence (how they might get lost playing hide-and-seek at the mall).

How many parents have been shopping with their young child in a large store, only to have their child start playing this game? This happened to my friend Leslie while she was shopping in a Walmart. Leslie had her son later in life, after finally having a successful pregnancy via IVF, and, like all parents, she treasured every moment with her son, Calvin. Calvin was a playful, rambunctious three-year-old at the time of Walmart-gate. Typical of children his age, he loved to play hide-and-seek, and he initiated a game with his mom. At first it was all fun and games because Calvin hid close by, where he could still see his mom even if she couldn't see him. But then he got bolder and went farther, until she couldn't find him and he couldn't find her. In a flash, her worst nightmare became a reality. I don't know if she flapped about like those Canada geese, but she flew to the service desk and Walmart went on lockdown in search of missing Calvin. That is their policy. No one comes in or out until the child is found and returned to his or her parent. This is incredibly helpful in protecting children from predators—in this case, I do not mean mountain lions, but other humans.

Unfortunately, degus don't have a lockdown system when they decide to play hide-and-seek, which may be why they haven't been observed engaging in this type of play in the wild, only in the lab.

These clever, sociable rodents, related to guinea pigs and chinchillas, make their home in the highlands of Chile on the slopes of the Andes mountains. Young degus do play. In the wild, they stay together when they play, running and jumping around burrows aboveground. They have a highly sophisticated alarm-call system, but young degus need some time to understand what to do when they hear a particular call. If the threat is by air, a degu is supposed to do one thing, and if the threat is by ground, it does another. Raptors or foxes are always on the prowl, and distracted, playful youngsters unfamiliar with their territory who haven't mastered the alarm-call system can easily become disoriented. This makes them a vulnerable target, and playing becomes more risky. So they stay together.

In the lab, it's a different story. There, they do love to play hide-and-seek, hiding behind objects in their enclosure. But they are in a completely safe environment. No raptors. No owls. No foxes. I find it remarkable that degus are able to perceive the lab as a safe-enough space to broaden their play behaviors.

Unlike degus, at least those in the wild, our children aren't regularly preyed upon, so they don't necessarily have inhibition in the same way. Like other species, though, play has enormous benefits— physically, psychologically, and socially. For the degus, a game as simple as hide-and-seek, where they learn to hide from and evade each other, gives them practice at some valuable skills. Similarly, play is critical to human development, and understanding this should inform parents' decisions about discipline: play has to happen in a *safe* space—a distinction kids aren't always capable of making—but unstructured play is also essential, so it shouldn't be overregulated. Having a sense of your child's limited grasp of danger, and setting reasonable expectations and boundaries around play, including when and where it happens, will ensure that he or she receives the full benefit and stays safe in the process!

Another problem for young animals, including humans, is getting lost in a crowd. We love to take our children to zoos and museums.

These places offer wonderful learning opportunities, but they are also very crowded, and it can be easy to get separated. This can be problematic for animals, too. Wildebeest herds can number in the thousands. Wildebeest are a type of antelope, and like many other ungulates, calves struggle to stand during the minutes after birth. On average, they accomplish this within five to ten minutes![15] But they are by no means steady on their little legs; they tumble and stumble over things in their way. Their next challenge is to get moving to keep up with their moms and the herd. This is no small feat, as the herd can move rapidly—since wildebeest can max out at fifty miles per hour, and calves are running before they can stand balanced! Within twenty-four hours they are sturdy and capable of following their mothers.

Getting separated from their mothers is dangerous for young wildebeest, which is why they have a built-in stay-with-mom system. As we explored in Chapter 3, this behavior is called imprinting, and it takes wildebeest calves a few days to imprint on their mothers. A combination of smells, sights, and sounds helps solidify the relationship between mothers and calves so that they remain together during this critical period.

Because calves will imprint on whatever large object is closest, including lions, mothers make sure to keep their calves close. If a calf moves away, the mom follows. Thus mothers take great care to avoid separation, especially in the first days, when calves are most vulnerable. Often lions, cheetahs, or other carnivores target young calves, hoping to separate them from the safety of their mothers. Roughly one third of calves are killed this way. That alone tells you it's fairly easy for a calf to get lost in the crowd.

While young human children are not likely to be taken by a predator, they frequently get lost in herds of humans. There are many potential dangers, including not understanding that some adults are safe and others are not—sort of like a wildebeest calf's not understanding that just because the lion is the largest thing it sees nearby doesn't mean it's safe to approach.

I discussed this issue of losing track of one's kids in crowded places with my friend Mindy. Mindy comes from a close-knit family of four sisters, and when she was a child, her family would go to the beach in northern California for vacation every year. As is common at many beaches around the world, lots of kids may end up playing in a certain area loosely surrounded by their respective parents.

These kinds of impromptu babysitting situations involving a collective of adults are very similar to what happens with Galápagos sea lions. As their name implies, this gregarious species is found on the Galápagos archipelago, a place still on my personal bucket list to visit. A few days after giving birth, mothers leave their pups and head out to sea to get some much-needed food. Not all moms go hunting at the same time, so sometimes the pups that are left behind will be loosely "watched over," much like the unrelated kids on the beach who gather to play under the protection of a group of their parents.

When humans do this, there is almost an unspoken agreement among the adults to keep a general eye on all the kids. Mindy relayed to me an incident involving her sister Allison that occurred on one of those childhood beach vacations. Their mother specifically instructed Mindy and her older sisters to watch out for the youngest, three-year-old Allison. Older siblings can find this cumbersome, and, if there is a large age difference, they often don't want to play in the same way. Because three-year-olds have minds of their own, restricted to only their worldview, when Mindy and her older sister refused to play whatever game Allison wanted to play, Allison decided she would go play on her own. Thus, while the other three became engrossed in their own activities, Allison wandered off.

At some point, their mother inquired about Allison's whereabouts, and no one knew where she went. If you are a parent and you have ever lost your child for just a moment, you can imagine the sheer fear, panic, and desperation that swelled in their mother's heart. Since it happened at the beach, one of the first thoughts was that Allison had drowned. After hours of searching, Allison was found safe with a lifeguard enjoying an ice cream cone, blissfully unaware of the drama surrounding her disappearance. Her mother and sisters hugged her, and everyone went back to their spot on the beach to continue playing. It would have been a crucial mistake to punish Allison because it would only have served to make her frightened about returning to her family. Why? Because kids her age cannot comprehend the "don't wander off" boundary and therefore need to have that boundary constantly enforced by a parent (or a willing and able sibling!). Repetition is a parent's best friend. Children need consistent and frequent reminders of boundaries. Telling a child anything once and expecting him or her to comply is completely unrealistic.

Sometimes parents or children get distracted. It's common for older children to become completely engrossed in something and not hear their parent say it's time to go, only to look up and suddenly realize they are alone. Jane Goodall described a situation where this happened to a young chimpanzee female, Fifi. Fifi was part of the notable F-family chimpanzees of Gombe. Flo, Flint's mom (mentioned in Chapter 7), was also Fifi's mom. Flo was arguably one of the best mothers in the community. At a certain point in the youth of a chimpanzee, the mom stops carrying her young on her back, and there is an expectation that the child will keep up when it's time to move. On one occasion, Fifi, engrossed in some activity, didn't realize her mother had left the area. When she did, she began to whimper and call out in distress. Flo, who presumably assumed her daughter had been following her, began searching and calling for Fifi once she realized she was not with her. And so the young Fifi spent the night alone. Scared and anxious, she fortunately survived the night and was reunited with her equally distressed mother the next day. The reunion was exuberant, both mother and child overjoyed to be back together. After much vocalizing and hugging, the incident was put past them and they went about their day. Chances are pretty good that Fifi, while distressed, was not traumatized by her ordeal and forgot about it. As with Allison, all is well that ends well.

To deal with the potential pitfalls of losing track of their kid in a crowd, some parents use leashes or other tethers to keep their children free-roaming but attached. Whatever your opinion on these options, other animals, such as dolphin moms, don't have tethers to keep their young close. But dolphins do make use of a special skill to keep their calves close. As with humans and many other species, wandering away from their mom—or swimming away in this case—is very dangerous for young calves, threatening their survival. How many mothers and fathers have run after their toddlers as they zipped away erratically? Mother dolphins end up doing more or less the same thing. Atlantic spotted and bottlenose dolphins have been observed chasing down their young while vocalizing to runaway calves that are under the age of three (roughly equivalent to a seven- or eight-year-old human child). Once a dolphin is older than three, it is classified as a juvenile and is rarely corrected for swimming off. It is possible that at this age they are large enough, fast enough, and know the terrain well enough to handle themselves.

And it isn't just dolphin mothers who will discipline their own children. Older "teenage" dolphins, unrelated females, and males will step in and set a calf straight. Remember, when we talk about discipline, we're talking about teaching and correcting and setting social and behavioral boundaries. In this case, the correcting is about teaching calves not to swim erratically away from their moms.[16] The chase to reel in runaway calves can last about thirty seconds; because many adults can intervene, young dolphins usually correct their behavior and stop swimming away within that time frame. If they don't, they are faced with getting "caught," where others catch up with them. In this case, the calf may get a buzz from the pursuer, who sends out a pulsing sound that ricochets off of the calf in a process called echolocation. Dolphins who are echolocating tend to swim upside down, which researchers believe may help direct the echolocation toward the youngster. Although there is no way to tell for sure, I think it would be safe to assume this produces an uncomfortable sensation for the calf, and gets his or her attention. Not a tether, but close enough!

Least often as a method of discipline—and seemingly only if absolutely necessary—a dolphin parent will physically contact its young. Sometimes, a parent will catch the calf and bop the child with its rostrum, the protruding bit at the end of its face (often mistaken for a nose or a snout, but the dolphin's rostrum evolved from its jawbone). Rarer contact behaviors are the tail swipe and pinning the calf down, actions that carry an inherent risk of injury. Given the danger, it makes sense that they are used sparingly against young animals. Dolphins, in general, are rather aggressive with each other, but they reserve physical aggression for other adults. Dolphins intuitively behave in ways that demonstrate an understanding that there is no benefit gained from harming, injuring, or killing their own offspring or kin. Thus, to guard against accidentally injuring weaker young dolphins, they rely most heavily on gentle, consistent discipline that teaches and incrementally increases enforcement when they encounter resistance. This illustrates an important aspect of discipline and coevolution of parental and offspring behavior in other species—namely, that most animals have "low stakes" means of enforcing boundaries for their young. And they exhaust all of those options first. By the same token, offspring almost always choose to comply before disciplinary actions escalate to high-stakes physical punishment. It is a delicate dance between enforcing boundaries and testing them.

The Knee-Jerk No

We've now seen that there are cases where animal parents must teach their young what to do and not do. But there comes a time when animal parents become more and more permissive. This is crucial, because for young animals to succeed, they have to have opportunities to learn and explore. If you are a meerkat, being able to handle dangerous prey is a necessary skill. While human children often have to be reminded to chew their food properly, inexperienced meerkat pups have to learn how to subdue prey such as venomous scorpions.

Interestingly, adult meerkats have only been seen teaching young pups how to do this in response to the pups' begging calls.[17]

WILD LESSONS

* From humans to degus, youngsters love to play. The difference is, degus know to play those rough-and-tumble games close to home, lest they get chomped. Human parents need to balance letting children explore the full benefits of play with keeping them safe until they can discern proper limits for themselves.
* Because the world is a loud and crowded place, it can be surprisingly easy to lose track of a child. Fortunately, you can develop a relationship with your child where you can verbally enforce boundaries. In those cases where your child completely resists, a tether may be an effective tool until he or she learns.
* If you or your child get distracted and separated, be exuberant when you are reunited. Dolphins don't body-slam their calves and chimpanzees don't grab and smack their kids when they find them!
* Physical aggression is a high-stakes form of punishment rarely used by animal parents and should only be used to *protect* young children from more serious imminent harm (oncoming traffic in the middle of the street, for example), not as a form of discipline.

Basically, begging pups are asking if they can do something. Instead of just saying no because it is a dangerous thing to do, adult meerkats offer a series of small yeses that facilitate learning. For example, they provide more assistance to young ones and progressively less assistance as pups get older. They even bring their pups prey items, such as scorpions, to practice on. All of these yeses in their various stages are necessary for the youngsters to gain the skills they need quickly and efficiently.

For many parents, life is hectic, and those moments when kids are begging or asking if they can do something, try something, or learn something are often inconvenient and—let's face it—can delay an otherwise well-planned day. You're running late, getting everyone off to school, and this is the morning little Susie wants to try to tie her shoes all by herself. Or dress herself. Yet, some animal parents capitalize on such instances to teach their offspring. And make no mistake, the inconvenient timing is costly to them as well. Orca mothers are second to none when it comes to parenting. Males and females stay with their moms for life, and a male orca is *more* likely to die before the age of thirty if he loses his mother.[18] This is because moms assist their sons in fights against other males.

In some places, adult killer whales employ a hunting technique called stranding, where individual orcas intentionally beach themselves on shore. They do this because all those delicious seal pups are located near the shore. This hunting style is complex and risky, and even six-year-old orcas still need help despite having practiced for several years by that age. As you might imagine, adults are more successful and efficient when they can do this on their own without slowing down to teach. Teaching takes time and comes with a very real and measurable cost: less food. And yet, the adults assist naïve youngsters by moving them up and down the beach, and when the juveniles give it a go, the adults are there to lend a helping hand, or fin, and rescue them from a sticky situation. There's also the risk that, because of their efforts to retrieve a stranded juvenile, the adults may inadvertently beach themselves.

Not all orca parents take the time to teach their offspring. Perhaps they don't want the associated cost, but for those orca moms that do, their offspring learn how to hunt successfully this way almost a full year earlier than those that are not taught! This means that, yes, it is inconvenient, yes, it takes time, and yes, you might be late for work, but slowing down to allow your children to learn whatever skills they are trying to master when they express a need, desire, or interest in doing so ultimately leads to faster self-sufficiency.

Of course lots of young animals do things that are dangerous, and the consequences can be severe, including injury or death. The same is true for our children. However, there is a difference between keeping a watchful eye and hovering. The current phrase for this is "helicopter parenting," and there is increasing evidence that it is crippling our children.[19] What is helicopter parenting? It is basically a knee-jerk "no" reaction to a child's behavior. It stems from overcontrolling, overprotective, overly intrusive parenting that likely has its roots in the adult's own anxiety issues being projected onto his or her child. *No, don't do that, it is dangerous. No, that is too risky. It's not safe.* I remember, as a child, being told not to climb a tree, not even a little bit, because I might fall out of the tree. To this day I am apprehensive of climbing trees. I suspect my mother was afraid of climbing trees and transferred that fear to me. Granted, not all trees are safe to climb, but not all trees are dangerous, either. It is more effective to teach children which trees are safe to climb, and many parents do exactly that, and all the better if a tree house is included.

Adopting the apprehensions of one's parents is not limited to humans. Rhesus macaques are very widely distributed and, as previously mentioned, live in complex social groups composed of female matrilines and unrelated adult males. These groups are characterized by strong dominance hierarchies. Rhesus mothers have a lot to teach their offspring, and being wary of snakes is on that list. Because of their broad distribution across western India and Asia, populations coexist with a number of deadly snakes, including cobras, various *Bungarus*, such as kraits, and pythons.

Biologists have found that when young, naïve macaques observe their parents react fearfully to model snakes, those youngsters develop a fear of snakes without ever having the personal experience themselves.[20,21] And despite this generalized acquisition of fear, there is still a nuanced fear response that is also learned: Adjust fear depending on the posture of the snake. To a rhesus monkey, a coiled snake, a hissing snake, or one with its head raised is a greater threat than one that is partially obscured.[22] This learning to be afraid of something is extremely functional in this example. The condition of being prey for snakes necessitates learning to be afraid. An environment containing predatory snakes has led to the evolution of learning this fear.

Something very important to consider is that human parents don't even have to say no to instill fear. Just acting fearful sends strong signals to most kids, who have evolved—like their macaque comrades—to be very sensitive to the emotional state of their parents. And snakes, ironically, are a perfect example of this. Many people have snake phobias, and when tested for an innate fear response to snakes, infants do show variability as well. But studies have revealed that in the absence of a fearful parental voice, they will actually reach out to grasp at images of a moving snake or non-snake with equal frequency.[23] Further study has shown that this holds true even for toddlers, up to thirty-six months.[24]

Basically, kids like to grab at living things and treat hamsters, gerbils, snakes, and spiders all the same: To them, they're all super interesting! And although infants and young children direct more attention toward snakes, there is a lack of convincing evidence that this attention is prompted by fear. Furthermore, infants actually show *less* of a startle response to snakes.[25]

But what if a child cannot physically hear a parent say no or see the emotional response? Can they still learn to be afraid of what a parent fears? Research is now revealing that even in utero, baby rats will pick up on fear experienced by their mothers and learn to also be afraid. How? By the smell a mother gives off when she is afraid. Oh, you thought it was just a catchphrase? Nope, smelling fear is very real. Researchers call this the intergenerational transmission of emotional trauma.[26] Although this research was conducted on rats, it revealed something stunning. Pregnant mother rats that were conditioned to find the smell of peppermint unpleasant, and therefore developed a negative or distressing response to it, passed this fear response on to their pups within days of birth, before the pups could see or hear anything.

Human children, like other animals, will acquire fear of harmless things simply because their parents find them aversive. The key, then, is to be mindful of whether we are teaching our children to be afraid of things that matter or depriving kids of their own experiences. If we're depriving them needlessly, then we're retarding a healthy development of fear based on knowledge and familiarity with something that is actually frightening to them. Strangely, we transfer our phobias and traumas to our young children all the time and fail to teach them to be frightened of things that are actually dangerous.

But helicopter parenting isn't just about fear; it's also about over-involvement, about not cutting the cord, so to speak. When we look at college-aged kids, those with helicopter parents suffer psychologically from higher levels of anxiety, depression, and a warped sense of entitlement, and are overall less well-adjusted to major life transitions[27]—not to mention they also feel violated as individuals, which may lead to oversensitivity in their relationships to normal behaviors that they falsely perceive as controlling.[28]

As we saw in the previous chapter, knowing when your kids are ready to make advancements and disconnect a bit from you can be difficult to judge. Bird parents have to be sensitive to lots of steps in their chicks' development. How do mourning dove parents know when their babies are ready to fly? To feed on their own? They must determine when to say "yes," when to say "not yet," and when to say "no." Animal parents have to be on the lookout for signals from their offspring that communicate when they are ready to try, to learn, or to *do* things on their own. Animal parents are extremely attuned to these signals, and we should be, too.

It takes great effort, conscious thought, a deep understanding of the needs, skills, and abilities of each child, and a commitment to supporting their growth and development into autonomy while setting appropriate boundaries. Having a mom that fights your battles for you into adulthood may work for orcas, but we aren't orcas.

And lastly, there are two necessary points to make. First, for all you helicopter parents who want to hover over other people's children, just stop. Imagine yourself sitting at the community park, complete with a jungle gym, and your child has requested your help in climbing up to the slide. You, as the parent, decide, *Nope, not this time, let him try.* And then another adult intervenes and lifts your kid up to help, probably shooting you a dirty look for ignoring your own child. The thing is, there is a distinct difference between rushing to the aid of someone else's child who is in danger versus making a difficult task easier because of one's own knee-jerk impulse not to let a child struggle.

On the other hand, and this brings us to the second point, if you're a parent who wants to give free rein to your kids, recognize that in doing so you are sometimes putting other children at risk, especially if they don't have a parent who enforces any boundaries at all, including social ones. Your child is not your peer and requires some limits, and you have a responsibility to set these limits. As always, it is a question of balancing freedom with boundaries.

WILD LESSONS

* Be conscientious about your own responses to things to avoid conditioning your children to be afraid of what's harmless. Remember, they can smell if you are afraid.
* There are certain things kids need to learn to be wary of, such as hot stoves, getting into cars with strangers, running in the street, and, naturally, poisonous animals and food.
* Raising kids takes a lot of energy, so animal parents are keen to encourage their independence. In other words, don't show up to college class with your child. Please.
* Neurotic, anxious, depressed, dependent adult animals wouldn't survive very long in the wild. By hovering over your children every step of the way, you are crippling them. They may survive, but they will not thrive.

continues . . .

- Barring imminent danger, don't hover over other people's children. Respect the parent who wants to raise his or her child to become a self-sufficient, secure adult.
- There is a fine line between free rein and zero boundaries. Children still need limits as they learn and grow. For the sake of your fellow humans and their children, remember to be your child's parent first and his or her friend later.

When Things Go Terribly Wrong: Maladaptive Behavior

Like human parents, most animal parents are larger and stronger than their offspring. Being able to moderate aggression is incredibly important. The eruption of conflict between two lions vying for feeding position at a carcass usually happens among adults, and even then, the individuals show remarkable restraint.

When we look at rates of child abuse resulting in physical injury in the United States alone, the majority of perpetrators are the biological parents.[29] In 2009 alone there were almost 500,000 substantiated incidents of child abuse and neglect out of sixty-two million children under the age of fourteen. In 2015 the number of reported victims was 683,000, an increase of approximately 27 percent.[30] The biological parents were responsible in 80 percent of these cases. Rates of abuse among mothers and fathers were almost equal, defying the myth that the majority of violent parents are male. Although the nearly 500,000 confirmed incidents represent less than 1 percent of all children being physically abused severely enough to be reported to Child Protective Services, and is variable from year to year, the data presents a striking contrast to what we observe in other animals.

As I have already mentioned, the vast majority of other animals do not use physical aggression as punishment against their *own* children. That is not to say that it does not occur, that animal parents do not neglect, abandon, or even kill their offspring. They do.

Why would parents abandon their offspring? It all goes back to that parent-offspring conflict we discussed at the beginning of this book. Let's look at birds and clutch size as an example, since birds have been the most well-studied group of organisms when it comes to parent-offspring conflict.

Scientists have spent an enormous amount of time trying to understand, explain, and predict clutch size in birds. Most people may not know this, but tons of bird species are living life on the edge—the edge of starvation. That means bird parents must strike a delicate balance between getting enough to eat for themselves and feeding their chicks. If things look good and a large clutch is hatched, great—until the environment or food supply takes a turn for the worse. Now there are too many mouths to feed. Birds have several strategies for dealing with this, some of which I discussed in Chapter 6. Here, though, I want to focus on abandonment.

The common eider is a sea duck found throughout Europe, North America, and even Siberia. Because common eiders breed in the Arctic where temperatures can be chilly, they line their nests with the fluffy, warm down feathers from the female's breast. When it comes to parenting, eider ducks have a few strategies. A male and female can pair up to raise the chicks, or females can form a crèche and share the work of raising chicks. For a colonial nesting bird, they lay clutches that can be fairly large, normally somewhere between four to six eggs. Males may stay with the females the entire time or take off about two weeks into the incubation period, which lasts about twenty-five days. One study found that over 40 percent of the moms abandoned their chicks. That is an extremely high number, even for birds. So what gives? Are common eider mothers just uncaring, awful parents?

To understand why this happens in eiders, we have to look at the condition of the mothers. Researchers discovered that females that abandoned their nest were in poor body condition.[31, 32] The duration of incubation, combined with the fact that mother eiders do not eat during this time and must rely completely on fat reserves, creates a dilemma for some females. If a mother continues to care for her brood, she risks death, whereas if she abandons this brood, she may survive and live to raise future offspring. Essentially, if a mama eider stays, starves, and dies, so do her chicks. But if she abandons them, only the babies die. This aligns perfectly with the parent-offspring conflict outlined by Trivers. Desertion is one solution.

And it is one that has been, and continues to be, practiced by humans, in some cases for the same reason. One could argue that child abandonment is adaptive in certain circumstances. We know that the conditions under which it occurs in other species are predictable and make sense from an evolutionary perspective. And where these conditions occur in humans, we see the same thing happen. The difference is that today, we have laws that punish parents for relinquishing parental rights to their children, despite the fact that this is precisely what they are compelled to do when they are unable to care for them, or, in cases of unwed teenage mothers or victims of familial sexual abuse, when it is far too dangerous for the mother to reveal she has given birth.

Historically, many human cultures practiced infanticide. In some cases this was because when conditions degraded and other older children, or the mother, were at risk of death or starvation, one option was to leave the baby in the forest. In medieval times and all the way through to the 1930s here in the United States, surrendered children were considered the property of whoever took them in. They became slaves, indentured servants, or laborers. Adoption, which we will discuss in Chapter 9, came later.

The reality that, like the common eider, human parents faced with dire conditions might be incapable of providing adequate care for their children is why we have safe haven laws in some states, so that a parent can leave a newborn to become a ward of the state, relinquishing all rights and responsibility to the child without fear of legal prosecution. Unfortunately, many states classify the same act as a felony when it involves an older child. This makes no sense, especially since it leads to an increased likelihood of physical abuse. If a person feels desperate enough to relinquish rights to his or her child and is prevented from doing so, it is pretty easy to imagine why physical aggression against the child is likely to follow.

This is where we depart from the vast majority of other species and swim into the dark waters of maladaptive parenting behavior. From an evolutionary perspective, it is illogical to repeatedly physically assault your child and not also desire to terminate your parental investment. In other species, whether in case of mating relationships or parent-offspring relationships, physical aggression is often a signal that the relationship is *terminated*. For example, some bird species may use physically aggressive contact to communicate it is time to leave the nest. But, as previously mentioned, this is very purposeful, and the offspring (or mate) complies. Because of our long developmental times, human children cannot actually escape the violence perpetrated on them by their parents or caretakers. Thus, in humans, violence against one's children, without relinquishing parental responsibilities and rights, is baseless and self-serving. For this reason, I contend it is pathological, and a disruption of the normal parenting process.

Given the rarity of chronic physical abuse in other species, there is a paucity of data of such instances in animals. What little data we do have from animals comes principally from primates. Although early studies on primates revealed that you could induce abuse for a certain length of time under certain conditions (for example, by isolating the infant from its parent and raising it separately), social isolation in rhesus macaques has not translated into a better understanding of this phenomenon in humans. Not to mention, these early studies artificially created conditions that would promote abuse. By isolating an infant from its parent early on, the bonding process was disrupted, and it is unclear if, when later reunited, the mother macaque recognized that youngster as her infant. It is fairly common, as we have already seen, for adult rhesus macaques to beat up on infants, just not usually their own.

When it comes to humans, it is easy to think that parents who abuse their children are suffering from mental illness. Unfortunately, this pat answer has not withstood the scrutiny of science. Naturally some portion of abusive parents are mentally ill, but mental illness is not a reliable predictor of abuse. Nor are poverty or economic hardship. Abandonment and abuse are not correlated, meaning that parents who desert their children are not more likely to be abusive parents. Stress, however, is a critical predictor of abuse, and risk factors associated with increasing stress are where we find the causal links.

Thus, poverty can be one potential component. How? We know that poverty causes an enormous amount of stress for parents, much as poor conditions do for those common eiders. The lack of resources, the constant pressure, and the lack of assistance increase stress and the probability of child abuse occurring in the home. However, poverty is not the only influence. The cumulative risk model suggests that as the number of potential risk factors increases, the probability of child abuse correspondingly rises.[33] What are some of these variables? The Child Abuse Potential Inventory, or CAPI, lists 160 items. Here are a few:

- Parental stress
- Parent was abused as a child
- Parental satisfaction (on being a parent)
- Success controlling child's behavior
- Difficulty of child (related to the controlling factor above)
- Socioeconomic status (a resource-driven and stress-related predictor)
- Family size and space (another resource-driven predictor)

The more factors a parent ticks off on the list, the more likely a child from that home will experience abuse.

What happens in cases where we observe abuse in other species? While, again, it's relatively rare and not well studied, there is evidence that abuse does happen in a number of other primate species—including pigtail, Japanese, and rhesus macaques, vervets, and sooty mangabeys—at a rate of about 5 to 10 percent.[34] Interestingly, as with humans, abuse runs in families. However, even in these species, closer inspection reveals that abuse that resulted in injury was more often the result of maternal *protectiveness*, meaning the mom was trying to hold onto her infant and the infant resisted or broke free and ended up injured in the process. Other causes of abuse include much the same ones we observe as predictors in humans: stress or an unhealthy social environment. Thus, abuse may really be a maladaptive form of redirected aggression. If a parent is under stress that they cannot exert much control over (e.g., job-related stress), the "normal" inclination to take aggression out on the source (e.g., the boss) is thwarted because of the risk of losing the job if if he or she retaliates against the supervisor. This unhealthy social environment creates an enormous amount of psychological and physiological stress that must be released somehow. One option for reducing this stress load is redirecting aggression on someone more vulnerable. Even lower-ranking macaques do this. As they say, crap rolls downhill. But there are also intergenerational patterns of abuse to consider.

Comparing the behavior of abusive and non-abusive rhesus macaque mothers gives us even more insight into this phenomenon. Abusive mothers tend to control or restrain their infant through contact, are more aggressive individuals overall, reject their offspring more often, and had abusive mothers themselves. That is very similar to some of the variables that increase the likelihood of child abuse in humans.

We know that intergenerational abuse happens. Research shows that children of parents who believe in and use corporal punishment think it is perfectly okay to resolve their own conflicts through aggression and use of physical force. The greater the frequency of spankings a child experiences, the greater the likelihood he or she will spank his or her own children.

Before we blame television and video games for the violence in the world, we need to take a hard look at how violently we raise our children. I think it's fair to say that the violence all around us is in some part a reflection of the violence we perpetrate on our children. We need to examine why we behave this way. Have we created such a stressful existence for ourselves that it is affecting our ability to parent our children properly? And let's be clear, we are not talking about physical force for protective reasons that lead to injury. We are talking about aggressive physical correction.[35]

I have seen firsthand the transfer of violence across generations. Many years ago, I had a friend who had a four-year-old son. The boy's mother regularly disciplined him with spanking. She was always financially stressed, and it wasn't clear that she actually wanted to be a parent at that time in her life, which translated into lack of patience and use of physical aggression against her son. One evening her son had a tantrum and hit her. She shrieked, "Where did you learn to hit like that?" He replied, "You!" with a level of contempt I did not know a four-year-old could possess. Sadly, like his mother, this now young man is aggressive and vicious in his arguments with others. I have little doubt that he will hit his children, too.

As if continuing the cycle of physical aggression isn't enough, brain-imaging studies of young adults on the receiving end of harsh physical punishment reveal that they have less gray matter in their prefrontal cortex.[36] The prefrontal cortex, as we have discussed, is incredibly important in cognitive performance and language development. But what constitutes harsh punishment? It could be anything from hitting in anger to causing serious injury. How negatively such behavior impacts an individual depends largely on his or her personality, temperament, and sensitivity. And for those of you who hold deeply to the right to hit your child, the world is increasingly saying "no" to this behavior. Forty-two countries have banned physical violence against children. The United Nations defines corporal punishment as "any punishment in which physical force is used and intended to cause some degree of pain or discomfort, however slight," and it calls physical punishment "invariably degrading."[37]

As I close this chapter, I want to make special mention of verbal and psychological abuse, since this can be even more common and equally degrading. If you are patting yourself on the back, saying, *Well, at least I don't hit my kid,* also take a moment to consider how you talk to your child. Parents who hurl insults at their children, scream at them, put them down, humiliate them publicly or privately, or practice other acts of psychological abuse are damaging their children just the same.

The absence of physical contact does not preclude a negative, harmful interaction. Emotional or psychological abuse is detrimental and destructive, and it impairs children in all areas of development.[38] But what constitutes mistreatment? This has been a difficult question to answer. Does it apply only if the intention is to emotionally or psychologically harm your child? No. Many parents will claim, legitimately and earnestly, that they love their children and will do anything for them while degrading and humiliating them in the very next breath.

A former colleague of mine comes to mind. Her son is the epitome of adorable. However, she often makes remarks about what a negative experience it was having him and how she would never have another child. It doesn't seem like she regrets having him, per se, but that is not the same as fully embracing and enjoying parenthood. She will fight like a mother bear to protect her son, even from the angry words of a stranger, but then turn around and spurn him or reject him by withholding positive emotional responses toward him. This, too, counts as ill treatment. Is she intending to damage her child emotionally or psychologically? Most likely no. But her behavior is even more disturbing in light of the fact that her son is very sensitive, doesn't contend with change easily, and experiences anxiety. She compounds this with her hostility toward these aspects of his temperament and personality. I fear that her harshness will have lifelong consequences for him.

And abuse doesn't necessarily come in the form of words. Giving a child the silent treatment, using a child to fulfill one's own psychological needs, holding a child to unrealistic expectations of behavior, projecting negative attributes onto a child (for example, "You're mean and selfish, just like your father!")—these all count as abuse, too.

Children can be difficult, but whatever their temperament or behavior, this does not absolve parents from the responsibility of ensuring their success in every possible way. Now I can't say for sure, but I'm fairly confident that cardinal parents don't scream at their chicks, *What the hell is the matter with you? Stop being a baby and just fly already!*

We humans think highly of ourselves when it comes to language—complex communication is one of our defining features—yet all too often we dismiss the negative impact our words can have on others. We believe that we can say things with impunity, without real consequence or damage to another individual or to our relationships with others. This is gravely misguided, especially where a child is concerned. A child is a person, vulnerable and dependent, with individual thoughts, feelings, and perceptions. Part of a parent's role is to recognize, acknowledge, and respect the individuality of his or her child. The failure to do so is a violation of the parent-offspring relationship. We may be the only species that has such a sophisticated command of the spoken word, but we are also the only species that uses this ability to reject, denigrate, humiliate, and harm our children and others in this way.

In an age when we take for granted that "we can have it all," no one likes to say this out loud, so I will: When we become parents, it would behoove us to fully embrace the parenting role and deliberately choose to enjoy that experience, because successful parenting demands that we prioritize our children's mental, emotional, and physical needs over our own, for a long time. If we are not prepared to do that, we will experience more than our fair share of frustration and resentment, which will set us up for greater risk of behavior that will harm our children.

WILD LESSONS

* Child abandonment and neglect in humans mirrors what we often see in other species: Environmental factors such as lack of resources increase the likelihood of abandonment.
* By providing a safe means to legally relinquish rights to children, we can offer desperate parents an alternative to inadequate parenting and reduce the likelihood of abuse.
* A parent's mental illness is not a reliable predictor of child abuse in humans.
* As for rhesus macaques, past history of abuse and stress (social and/or financial stress) are strong predictors that a parent will abuse his or her child.
* Physical punishment basically leads to brain damage.
* The United States is behind forty-two other countries in the world in not banning the use of corporal punishment against children.

CHAPTER 9

Different Families

When I was growing up, I had the sense that my family wasn't normal compared to the families of the majority of my peers. My parents divorced when I was six and my father was gone. Vanished. The exception was Melissa's family. Melissa was my best friend from childhood; I met her when my family first moved to the United States from Italy. She may very well have been the first friend I ever made here. We are still "besties" to this day, and her mom is like a surrogate mom. Melissa was the only one I knew who also had divorced parents and, as mine did, her tiny grandmother (she really was tiny!) lived with them. I was always trying to wiggle my way into other families, and because Melissa lived close by, I spent a lot of time at her home. Melissa and I were always having some adventure and running into one-of-a-kind characters. There was the frightening and creepy neighborhood flasher, the older boy who swept Melissa off her young feet, and the strange Chinese boy who tried to kiss me like a Komodo dragon. Unlike in my family, though, her mother never remarried, her grandmother was a permanent fixture, and her dad was still very involved in her life—she saw him regularly.

After my parents divorced, the man my mother married entered the picture with two children of his own, a daughter and a son. They didn't live with us full-time, but instead visited every other weekend and during holidays. That meant that, as a kid, I had no biological father present, an unrelated man living in the house, and periodic visits by his children—"siblings" with whom I had no biological connection. And this was all during a time when the concept of the ideal American family was something much more fixed in the minds of most Americans.

For some years now, we've been hearing about the breakdown of the nuclear family—typically composed of a mom, a dad, and children—and the loss of "traditional" family values. Since 2010, the US Census has revealed that when we look across America (and, really, the world), there is a lot more variation out there in what constitutes a family. And that diversity is growing. As I mentioned in Chapter 4, mothers in particular are criticized for working. Similarly, when we discuss single parents, we seem to focus all of our attention on single mothers. Are these new female roles really prompting a crisis, though? With the picture of families changing, are we doing something out of character that is actually harmful to our children? When we look at other species, is it always this isolated nuclear family unit that's coveted so deeply? And what about divorce? We know that divorce happens in other species where males and females pair for life. Sometimes it doesn't work out for them, either. What happens to those kids when their animal parents separate?

Is Divorce Really All That Bad?

The one-size-fits-all model of family and the traditional approach to parenting are clearly not working for us. This is not all that surprising since, historically, it never really has been this way for humans. Across cultures, and indeed throughout human evolution, there are many iterations of the concept of family. Only recently have we, as a society, seemed particularly hung up on the mother-father-children model of family. Let's tackle that one first.

Strict biparental care—with no assistance from older children, relatives, or nonrelatives—is relatively uncommon when we look at other species. One exception is the prairie vole. Prairie voles are touted as the model animal for understanding human relationships. Why? This little rodent, which seems sort of like a hamster but is more closely related to a lemming, has all the attributes we idolize: monogamy, parental devotion (from both the mom and the dad), and grief when a partner dies. Family life in prairie voles is a well-oiled machine, driven by many of the hormones discussed in Chapter 2. It certainly does seem ideal. Both the mother and the father attend to the nest, coordinating care for the pups so they are not exposed prematurely to the elements, and often huddling together with their children.

The Peruvian poison dart frog, seemingly an unlikely candidate for the model family, also displays behavior associated with our familial ideal. These frogs come in all varieties, including ones that look like they are wearing a pair of blue jeans on their legs. In particular, mimic poison dart frogs have been the subject of much interest because these tiny, colorful frogs enter into long courtship periods (you know, for frogs). Once they pair up, they are faithful to each other, and both parents provide care for all their baby tadpoles. After settling on an area to live and lay the eggs (in leaves away from water), both the mom and the dad work to defend the home range and protect their clutch of soon-to-be tadpoles.

Parental care by the mimic poison dart frog is a herculean task. Once the tadpoles hatch, the dad carries each of his kids, individually, on his back, to their own small pool. He cares for them while they develop over the next few months. Now you may be wondering, where the heck is the mother? She is out foraging so that she can make unfertilized eggs to feed all the baby tadpoles. Because their dad is so attentive, he knows when the tadpoles are hungry and calls out to their mom to come deliver an egg to this or that tadpole. So how did idyllic family life evolve in this small, toxic amphibian? Researchers think that using small pools of water for each tadpole, rather than a considerably larger water hole, leaves little for the tadpoles to eat. The solution: The mom and the dad work together to feed the kids. What happens when you take either parent away? About half of the tadpoles die.[1]

This is not terribly uncommon. In species that require both parents, when the parents split, or one parent deserts, the kids suffer. This is largely because the other parent cannot compensate for the absent parent, and survival of all the offspring necessitates the participation of the other parent. In the case of mimic poison dart frogs, single parenting is not entirely catastrophic, but I think we can agree that half your children dying is a fairly poor success rate. Although half of all human children don't die after a divorce or desertion by one parent, we certainly hear about how detrimental it is to the emotional, mental, and physical well-being of children, as well as how it can fracture their development of healthy, functional relationships in the future.

This perception that children of divorced parents suffer psychological and developmental damage, going on to becomes less successful adults, stems principally from US-based research outcomes and the agenda promoting the family deficit model. The deficit model was based on the underlying assumption that children of divorce are inherently worse off and asserted that a two-parent family was necessary to ensure children's successful socialization and development and that in its absence children would be psychologically damaged.[2]

Marital insecurity has increased globally, bringing additional focus onto this matter, and not surprisingly, the early American theories that children of divorce are worse off simply because their parents split up have remained largely unsupported as other countries have picked up the ball and begun to look more closely. Specifically, a German-based study found a fairly trivial effect on adolescent well-being and very limited disadvantages for children of divorce.[3] Two of the biggest factors that may actually cause a problem, however, are financial strain and/or stress caused by the parents during their separation.

Interestingly, the model family unit of prairie voles is not as rigid as the one we idealize for ourselves. Sure, the majority of prairie voles may conform to the "golden age of marriage" version of family, but there is a ton of variation. In addition to the mom-and-dad version, you can find single moms, single dads, and large communal families of prairie voles. What happens to the pups if they lose a parent? Not a whole lot. When single moms were examined, it was discovered that even though they didn't pick up all the slack (which did mean that pups experienced a bit more exposure to the elements), the pups were cared for just fine. It would seem then that divorce is much worse for mimic poison dart frogs than for prairie voles and human children! One interesting difference is that single-mother prairie voles have less time to engage in "fun" activities, such as explore, wander around, jump, dig, or otherwise goof off. [4] Many a single parent can relate to the all-work, no-play lifestyle.

Is this the full picture, though? Is there ever a time when it is actually *better* to have only one good parent instead of two parents if one of them is of poor quality? Since animals don't remain in partnerships that aren't working, we can't evaluate whether staying with a bad parent or kicking the partner out and taking over all the parenting duties works better for them. However, we can look at what happens when a parent is dealing with a substandard partner, and whether the competent parent does better alone. To test this in animals, one research approach has been to remove a parent and figure out if the remaining one can match or provide extra care. Another approach has been to handicap one parent, making him or her *less* able to provide care. Because a large number of birds rely on a two-parent model of family, most of this work has been done in birds, but for our immediate purposes, let's turn our attention to burying beetles. Yes, burying beetles.

As their name implies, these beetles bury stuff. Not just anything though. They will bury the carcasses of other animals, such as birds, mammals, or other dead creatures that will serve as food for their babies. They don't bury things any which way, either. When a mom and dad come across a potential food item, they work together to dig a hole beneath the animal, remove all the hair or feathers, cover the body with antimicrobial and antifungal secretions, make a ball out of the body, and repurpose the hair or feathers to line the chamber. The female then lays the eggs near the chamber. This whole process takes a fair amount of time, sometimes upward of eight hours.

Despite all of this early prep work, the parents aren't finished. After the larvae hatch, both parents supplementally feed their hungry brood by regurgitating liquefied bits of the carcass. They keep up parenting duties for several days until the offspring can fend for themselves. But what happens if one parent is a slacker? To investigate this, scientists glued tiny little weights, the equivalent of half their body weight, on these tiny beetles, effectively slowing them down a bit. They then compared who did what for how long between pairs that had no weights and pairs where either the male or the female were bogged down with extra bits of aluminum. (I know, we scientists can study some strange things.)

What did they discover? Once the extra weight was added, both mother and father burying beetles increased their efforts to make up some of the difference that would otherwise be provided by their partners, but it was not fully compensatory.[5] But what if the mother or the father beetle ran off? For the most part, the parent left behind made up all the difference and actually provided more care than both parents had as a pair![6]

Ironically, what we see in many birds and burying beetles mirrors what I think we see in humans: A single parent tends to compensate more for an absent second parent than for one who is not living up to his or her full parenting potential.[7] It's noteworthy that animal partners don't stay together, bickering and fighting. There is no domestic violence or abuse in monogamous pair-bonded animal partnerships. There is pretty strong evidence in humans that children living in two-parent households with high conflict have *more* problems psychologically than those in low-conflict divorced families.[8] And it doesn't have to be dramatic conflict, and the conflict need not involve the children directly.

Stress is bad for kids, and experiencing conflict in the home or during a divorce is the biggest problem for children. So maybe "staying together for the kids" isn't the best option, since this often leads to worse conditions. Instead, it may be more beneficial for fighting parents to separate amicably. However, it seems that, unlike burying beetles, when we divorce, our ability to parent diminishes as well. In the case of humans, I suspect the cause has more to do with getting wrapped up in the conflict of divorce, resisting separation, and power struggles between the parents that throw the children in the middle, leaving little time and energy to take care of their kids. We would do well to understand that divorce and separation can often be the best choice, and to begin learning how to handle it more appropriately to minimize the impact on our children.

WILD LESSONS

* The "golden age of marriage" was but a cultural blip. The reality is that even among other traditionally monogamous nuclear-family species, there is more than meets the eye. Unless you are a mimic poison dart frog, in which case you really do need your partner.
* Divorce, separation, desertion, infidelity—all of it happens in animal families, too. Providing enough resources as a suddenly single parent—animal or human—is one of the greatest obstacles to parenting success.
* When a barn owl divorces, it goes on to have a better life, usually a better next partner, and the owl's offspring fare better, too. Staying together "for the kid's sake" is hardly optimal when there is continuous or intense conflict between the parents.
* Kids and burying beetles do better with a single parent rather than two when one doesn't step up properly to the task of raising them.

> * When we humans separate, we need to learn how to separate amicably! Life is hard enough without adding to the pain and conflict. Other species don't tend to have nasty separations.

The Single Parent

As we know, sometimes single parenthood is forced upon a mother or father, while other times it is a choice. Typically when we think of single parents, we automatically think of a single mother. Single motherhood is seen as a social crisis, while single fatherhood is perceived as saintly. The reality is somewhere in between, where there are no saints or sinners, merely single parents trying to raise their children successfully. Many animals, both male and female, raise their kids alone. And make no mistake: It is hard work, especially for those species for whom two parents are ideal. Even though one good parent is sometimes better than two bad ones, or a pair riddled with conflict, or even one good one that stays with a poor-quality partner, there are some unique challenges with which single parents must contend.

One of the biggest challenges is dealing with simple logistics. If you are a single mother or father, it is likely that you are also working. And if you are like a cheetah, you don't have help. Although cheetahs may be best known for their superior speed, to me a cheetah mother always represents the iconic single parent. As I mentioned, I met my first cheetah in the late 1990s at the Miami Metro Zoo. He was the first cheetah they received, and I was able to visit with this one-year-old beauty. I'll never forget seeing firsthand and up close how, like a football player who wears eye black to reduce the glare of the sun, a cheetah has a natural black streak lining its face from the corners of the eyes down the sides of the nose, helping the animal hunt its prey.

Single females can have as many as five cubs at a time, though in such cases it is rare that all cubs survive. Three cubs at a time seems to be about the average, and that is more than enough for cheetah moms to contend with. Before giving birth, a female will find a suitable location, or lair, in which to have her cubs. Here the cubs will stay for three to five weeks until they are more mobile.

During this time the cubs are especially vulnerable, as they are essentially being left home alone, which is dangerous. Leopards, lions, and hyenas may discover the lair, and all three will kill cubs that they find. Yet the mother cheetah must go to work. Occasionally other disastrous random events occur, such as fires, and these baby cheetahs, who cannot walk yet, will burn or otherwise meet their death. But the energetic demands of nursing means that the mother cheetah cannot afford to stay at home. If she doesn't eat, her cubs will die anyway.[9]

For many human parents, too, if they don't go to work, their kids may go hungry and their families may become homeless. In the United States, there are only a few states that have laws regarding the minimum age a child can be left home alone, and it ranges from eight to fourteen years old. That's a heck of a spread. When I asked my friend Alma how old her daughter was when she first left her home alone, she said nine. I was a nine-year-old "latchkey kid" myself. Clearly, newborns and young children should not be left home alone like cheetah cubs, but what is the "right age"? I'm not sure that there is one right answer to this question, but I do think that we need to have a better support structure in place to help parents (1) decide when it is appropriate to leave a particular child home on his or her own and (2) gain easier access to high-quality supervision until then.

Naturally, these issues disproportionally impact low-income single parents, as many of them cannot afford daycare or after-school sports activities. Such was the life of a longtime single-mom friend of mine, Gina. Gina was no sleek, calm, regal cheetah mom; she was as frantic as a hummingbird, always on the edge. Hummingbirds are virtually on the edge of starvation every single day . . . maybe that is why they are so darn cranky! Gina was frequently on the edge of eviction, the cusp of no electricity, and sometimes she just plumb fell off the ledge.

I met Gina while working a brief stint at one of those labs that processes medical samples. A few weeks of working the overnight shift was enough to send me back to waiting tables. There was no way I could garner enough enthusiasm for a brutal third shift at eight dollars an hour. Making a hundred and fifty bucks a night working the swanky Japanese sushi bars was the only sensible option. By the time I hightailed it out of there, Gina and I had already become fast friends. A few years later, she found herself pregnant and alone.

The next fourteen years of her life were one catastrophe after another: evicted three times for not being able to pay rent, fired for missing too many days of work (due to a sick kid), having the electricity or phone cut off too many times to count, and raising her son on a diet of fast food because it was all she could afford. I distinctly remember one time when, unable to pick her son up from day care on time yet again, the day-care center flat out refused to allow her to bring her son there any longer. This was pretty disastrous. Ultimately, Gina had to bring him to another facility that was open later but charged more after 6:00 PM and charged *by the minute* if parents were late.

In the end, Gina tapped out of the single-parenting game and sent her teenage son to live with his much older brother (her son from a marriage years earlier). Having older siblings help raise younger ones isn't a human novelty. It happens in other species as well—though, to be fair, the comparison isn't quite as direct. Rather than sending their chicks off to live somewhere else with another offspring, older red-cockaded woodpecker sons, for example, either stay at home or move next door and help raise the next generation.

Red-cockaded woodpeckers are common to the southeast United States, and like other woodpeckers, they feed on insects and sometimes fruits, and they nest in tree cavities. A male-and-female pair will maintain and defend territories year-round, and it's almost always the sons that stick around to help their parents. This happens in 30 percent of red-cockaded woodpecker families. Why they do this is an interesting question, especially since the parents usually remain together, though occasionally, as in the case of Gina, the female moves on and finds a new male, and so the helper that stays behind is really a half brother helping his dad out. At any rate, it seems that this benefits not just the parents and the next round of offspring, but the sons who are helping as well.

So, if you are an older child, why help? One reason is that moving away from home can be dangerous, especially for males who must compete with other males for space. In red-cockaded woodpeckers, males that stay near home don't die as often.[10] So, even though helper sons help build nest cavities, incubate eggs, feed their full or half siblings, and help defend the territory, all of which leads to greater success and survival for the nest,[11] ultimately it's selfish. I'm not sure the same could be said for Gina's older son. He didn't gain a measurable benefit from having to take over all the emotional, mental, and physical costs associated with being responsible for a teenage half brother.

The generalized term for having helpers in animal societies is "cooperative breeding." We will explore this in more detail later in this chapter, but it is worth introducing now. Sometimes, as with Gina's older son and red-cockaded woodpeckers, these helpers are relatives, so it seems easier to explain, but other times, as in the case of warthogs, helpers aren't related at all.

On the surface, warthogs seem belligerent and constantly on the offensive. They are scrappy, that's for sure. But they have to be. They are constantly contending with imminent danger. Belonging to the pig family, they are armed with a serious set of tusks. They use these in arguments with each other and to fight off predators, though their first strategy is to run. And when they are on the move, warthogs are fast, zipping here or there as if perpetually on a caffeine buzz. When it comes to social structure, warthogs live in mixed yearling (or "teenage") groups or adult-only female groups. The males are generally on their own, but will sometimes form bachelor groups. Some females also live and raise their kids alone. For those that stay in a group, however, there are some perks.

Babysitting is one of the benefits of having a group of other single parents around. In the case of warthogs, females initially give birth to their piglets alone, but later combine their kids into a single burrow. This is different from having family to help. Oftentimes these females are unrelated, but they still watch out for the kids of other mother warthogs.[12] This approach, having a network of other single parents available to assist, may make the difference in survival and success for some warthog piglets.

In humans, there is some evidence to suggest that single individuals maintain closer ties to family and have a broader social network, from neighbors to coworkers, and are more willing to offer assistance to others. For married couples, once children are on the scene, the research shows the numbers of social ties go down due to the energetic demands of parenthood,[13] and the same is likely true for single parents. This means that for single parents it can be even more crucial to develop a strong social network with other single parents in order to gain the benefits of cooperative support, something that was sorely lacking for Gina.

Although the lack of community assistance and support for single parents isn't a given in every community, culture, or country, the current societal structure in the United States does not lend itself to helping single parents deal with some of the logistics that come with working and raising kids.

Getting assistance doesn't seem to be as hard for single fathers as it is for single mothers. As often happens when I am writing about one topic or another, I go out and talk to people to round out my research. This is how I came to be acquainted with Tom. I was very interested to hear about what it is like to raise a child alone as a single father. Now, I have many male friends who have been active, fully participating fathers after separating from their partners, but I was keenly interested in getting the perspective from a father who was the sole parent.

I met Tom while staying with friends in Arizona. Going to the desert in the height of summer was not the most brilliant idea I ever had. One day, as temperatures soared to 114 degrees Fahrenheit, the air conditioner broke. Tom, the repairman, came to the house and, because I am constantly asking people questions about their lives, I discovered that he was a single dad to a now fifteen-year-old teenage daughter. When he happened to also mention that he had cared for her single-handedly since she was an infant, I was intrigued.

His story calls to mind the fact that when we think of parenting in other species, we have a tendency to focus on the moms. But there are many species where dads are the sole caretakers of the kids. Rheas are a prime example of dads in charge. Related to emus and ostriches, these large, flightless birds sometimes get a bad rap for the small size of their brains. But what they lack in brain volume they make up for in spirit. The greater rhea is found in South America, and males definitely like to play the field. They mate with several females, and each one deposits eggs into his nest before moving on. A single male may end up with as many as fifty eggs at one time, though his average is closer to half that amount. Aside from building a nest and incubating the eggs, rhea dads are solely responsible for raising the chicks and teaching them everything they need to know about being rheas.

The incubation time depends on how many soon-to-be chicks are in the nest, but it's approximately forty-two days, and males spend basically all of that time on the nest except for short periods of eating nearby. After his chicks hatch, a rhea dad provides exclusive care for the next four to six months. His duties range from herding all those chicks and keeping the family together to watching out for predators and teaching the chicks what to eat. As seen in warthogs, sometimes males will fuse their brood of rhea chicks together, or a male will adopt other chicks to add to his family.[14] But more on adoption later.

The interesting thing about species where males incubate eggs, or handle all the pre-birth duties and post-birth tasks, is that a male with a nest is sometimes a chick magnet, in the female chick sense. This is especially true for many species of fish where females will preferentially lay their eggs with a male that already has eggs.

Right off the bat, as Tom described to me what it was like to be the sole parent to a child, he remarked that he was flooded by offers of help by women. He said it was like having a puppy. I told him it was more like he was a sand goby! The offers of help were automatic and instantaneous. He didn't have to work too hard for them. In his case, his ex-wife was sent to prison when his daughter was two weeks old, and he was thrust into being responsible for all her needs. He confided that he was completely clueless about what to feed her, what diapers to buy, and every other detail imaginable. He and a pal headed off to the supermarket, newborn in tow, and stood mystified in the baby aisle until a woman appeared. Immediately they asked her what babies eat. He told me that, at first, the woman was suspicious of these clueless men holding a newborn, but once she realized they hadn't kidnapped her, she sprang into action, filling their cart with everything they needed and offering instructions on how to take care of the basics.

The pattern of easily acquired assistance from women was repeated throughout the next fifteen years and proved helpful as his daughter grew older and began needing more input on female matters such as hair, makeup, boys, menstruation, and training bras. So why is it that single dads attract unrelated females to help, but the reverse isn't as common? Single mothers definitely don't have men lining up offering to help raise another man's child. By and large, we see that males of other species, like human males, tend to only provide parental care to offspring they have some degree of certainty are theirs.

As we've mentioned throughout this book, parenting is an enormous investment, and it doesn't make a lot of biological sense, from a natural selection perspective, to provide all that care to offspring that aren't even yours. Females, on the other hand, may see a male that has offspring as *more* attractive—not necessarily because they want to raise someone else's children, but simply because if another female chose him to be a father, he must be a quality male.

I mentioned above that male fish benefit from this behavior quite literally, and that Tom was the equivalent of a sand goby to women. Let me explain. There are many varieties of goby, and they are a rather interesting group of fish for reasons that go well beyond parenting. For instance, they can change their gender in just a few days when conditions are favorable. But we are talking about parenting, not transgender fish. So, when it's time for sand gobies—small, bottom-dwelling marine fish found off the coastlines of Europe—to mate, males build a nest in the sand (hence their name) under a clamshell or mussel shell. They then use the sand they dig up to cover the shell to camouflage the area. When it comes to male sex appeal, having not just a nest, but one with eggs already in it, makes you a real contender. In gobies, males tend to be the sole provider of parental care; they protect the eggs, fan water over the eggs to keep oxygenation levels up, and clean the nest during the weeks leading up to the eggs hatching. It turns out that males with eggs already in the nest appear to be better fathers. The more eggs they have, the better they care for them, so females prefer males with kids.[15]

In Tom's case, he quickly began to notice that it was principally women with no children yet of their own who wanted to quickly assimilate into his little family unit. He became increasingly selective about which women he allowed into his daughter's life and ended up with two consecutive long-term relationships with women who had both already had children that were grown. Thus, he got the benefit of experienced mothers who were not looking to be his daughter's mom. Whether you are a single mother or a single father, it can be incredibly important for your child's well-being to have great discernment as to when and if to allow another adult into the picture.

WILD LESSONS

* Childcare is a challenge for all working parents, but disproportionately for single parents, particularly those with limited incomes. Leaving children on their own is an

age-dependent decision for other species as well. We may not reach consensus on the "right age" to leave a child home alone, but we can provide better support to working parents to assist them when they face these difficult circumstances.

* If sending your younger child off to live with an older one isn't an option, another option for single parents is to form a solid network of support, warthog-style.

* On average, single mothers tend to receive less help than single fathers. Making resources and assistance available to all parents (free day care, parenting classes, support groups, instruction on basic care, etc.) benefits all children.

The Stepfamily

For reasons discussed briefly above, having nonbiological parents step in and assist in raising offspring that isn't theirs is rare in other species. Even though stepfamilies are increasingly common for humans, they present serious dangers and pitfalls that affect our children. This is not to say that things can't run smoothly for such families, but more often than not there are problems that crop up. To try to get a handle on this, let's start by picking up the thread we ended with in the previous section.

The enormous costs of parenting are a given. Let's face it: Parenting is a mammoth task when it comes to time, energy, and resources. It's plain expensive, and from a purely biological perspective, it doesn't necessarily pay to spend all that capital on someone else's child. This is one of the primary reasons we see infanticide in other species. Lions are perhaps the classic example. Social life for lions revolves primarily around the female. They are all usually related—sisters, mothers, daughters, cousins, aunts, etc. The males come and go depending on how long they can maintain their status against other males. The average length of stay for a single male, if he can even successfully defend a group of females against other males, is less than two years. The same is true for pairs of males, though they start off with a higher success rate. As the number of males teaming up to lead a pride increases, so, too, does the length of time that they can hang onto the pride, ranging from three to six years![16]

When new males take over, they kick out all "teenage" males and females (aged two to four years old) and kill young cubs. They kick out older kids because they don't want interference from young males. Females don't reproduce until they are at least four, and they only have cubs once every three and a half years. This is shorter than the tenure of most males, which means new males coming in won't get to mate with all of the females. By killing the younger cubs, they induce the females to go back into estrus. Essentially, stepparenting is nonexistent in lions. However, female lions don't just sit back and accept this. They gang up, helping the resident male(s) fight off intruders. If the existing males are successfully chased off, then the females still battle to save their cubs. In the end, though, they fail.

Infanticide is a huge problem for animal parents, and it isn't just males that are a threat. Recent research is showing that females, not males, are the most frequent perpetrators. Replacement females—or to put it in human terms, the second wives—are the principal aggressors. House sparrows may seem like innocuous little birds. And to us, they are, living in our backyards, taking advantage of all the food we humans provide to wild bird populations. But among each other, out there in the sparrow world, it is a different story, a regular soap opera.

Like other social birds, house sparrows can hang out with each other, sharing dust or water baths and flocking together to sing or eat. They also nest communally, preferring nest boxes clumped together. However, this preference sets the stage for a lot of conflict. Normally, pairs mate monogamously and raise their offspring together. Sometimes, though, the pair breaks up, or another female joins the nest. When that happens, the new partner will attack and often push the unhatched eggs out of the nest or kill the hatchlings from the previous mate.[17] What is interesting is that the reasons why male and female house sparrows commit infanticide are different. Males that kill off a brood have usually prompted divorce in an existing couple, and then the intruding male is quickly rewarded by mating with the female and having his own clutch of eggs to raise. This implies that, in house sparrows, males use infanticide as a mating strategy. This is very similar to what happens in other species. But females have an altogether different motivation. Infanticide usually happens when females have to share a mate. Basically, by attacking the chicks a male had with a *different* female, a female is able to secure most of the resources and attention from the male toward *her* offspring.[18]

We can see how this might parallel the situation in human divorced families, whether or not there are new children being added to the mix in the second marriage. Let me use what happened to my best childhood friend, Melissa, as an example, because it is a classic house-sparrow scenario representing the female side of the equation. As mentioned earlier, her father and mother divorced when she was young. Her father remarried while her mother did not. His new wife, Rebecca, had a son from a previous marriage, and Melissa's dad stepped in as a positive father figure to him. Unfortunately, the reverse was not true for Melissa. Now, granted, she did not need a mother figure—she had a fantastic mom—but having a positive relationship with another female adult could only have benefited Melissa. However, Rebecca was extremely jealous of Melissa's mother, and possessive over Melissa's father's attention, time, and especially his money. In true house-sparrow fashion, she attempted to undermine Melissa's relationship with her father and divert his focus exclusively toward herself and her son. Although in the end Rebecca was unsuccessful at severing the relationship between Melissa and her dad, she was able to generate conflict, stress, and angst between them for many years.

In humans we call this the Cinderella effect, though it applies to the dynamics present among males or females. When we look at the data in humans, many of these patterns emerge over and over again in stepfamilies. Setting aside any incidence of direct abuse or neglect, which we already know is committed at a higher rate by stepparents, a subtler effect is seen.

First, it has an effect on general attentiveness and vigilance. Whether we are talking about families with a single biological parent or two, watchfulness and risk of accidental injury is approximately equivalent. However, for two-parent families composed of a biological parent and a stepparent, the risk of injury and death is remarkably higher.[19] When researchers examined this more closely, they discovered that this was not a consequence of outright abuse, but rather of the tendency to be less watchful, which leads to a greater frequency of accidental fatalities when children are being supervised by a stepparent. I think parents generally realize others may not be as cautious with their children as they are, but might not recognize this extends to their new partner. And just to clarify, this lack of attentiveness is not intentional.

Another subtle difference in how stepparents treat children involves time and money. When researchers interviewed men about the amount of time and financial resources they spent on children, the amounts of financial investment reported were substantially different depending on whether the children in their current relationships were biologically related to them or not. The men also spent less time with children who were not biologically related.[20] I saw this firsthand in my own situation. My mother's new husband disproportionately provided for his own children from a previous marriage more than for my brother or me. We got their hand-me-down clothes after they were done with them. It was particularly noticeable at Christmas. While his children had piles of presents, my brother and I received just a few things, whatever our mother could afford.

This disparity crosses genders, meaning it's not just men who allocate time or resources differently. Melissa related a story to me about dinners at her father's house. When Melissa was visiting, her stepmother, Rebecca, would make a spectacular meal for Melissa's father, herself, and her own son, who was close in age to Melissa. What did Melissa get for dinner? Usually a plate of spaghetti with ketchup.

Sometimes the difference in care is conscious and sometimes it is inadvertent. Regardless, we know that disparities in treatment exist, and they can be downright hostile. It begs the question, why do biological parents, like Melissa's dad and my mother, accept their children being treated poorly? In my case, my mother's husband didn't start out treating my brother and me poorly. He put a lot of effort into appearing to be a good guy, happy to step in and raise two kids that weren't his. In animal behavior, and even in evolutionary psychology, we call this mating effort. Basically, to convince my mother to become his partner, he behaved in ways that would seal the deal. Happens all the time. Even in vervet monkeys.

Yes, the very same sandwich-stealing monkey I encountered in Kruger National Park is also a bit deceptive when it comes to how males interact with infants that aren't theirs. In an effort to court a female, a male vervet monkey will be friendly toward her kid. He may sit close, groom the kid, or play with him or her. Not surprisingly, females are more attracted to males that have positive interactions with their offspring. A clever experiment with a one-way mirror revealed that when some males *thought* the mother wasn't watching, their friendliness disappeared and they became aggressive to the infant or young vervet.[21] Sometimes we, like vervet monkeys, can be duped. The difference? When vervet mothers observed a male being hostile toward her offspring, she severed her relationship with him. Immediately.

In other species, stepparenting is relatively rare, but where it occurs you still see a spectrum from outright infanticide to indifference to complete adoption. In eastern bluebirds alone there is great variation, with one study revealing that only 30 percent of replacement males actually helped feed nestlings that were not theirs.[22] Similarly, in the case of humans, there is a greater difference in the amount of financial and temporal investment individual stepparents make compared to the variability found among biological parents.

This is not meant to be an assault on stepparents, but there is a reason for the stereotypes that we have, and we see unmistakable parallels when we examine the data on humans and other species regarding abuse, neglect, infanticide, indifference, and differential investment. So what can we do? We can, and must, be extremely cautious about who comes into the lives of our children. We can critically evaluate, independent of the adult relationship, how a potential mate behaves toward and treats our children. Furthermore, even though infanticide is less common for humans than for, say, lions, by tolerating abuse, indifference, or disparate treatment, our children are being damaged. As parents, we have a responsibility to defend and protect our children.

That does not mean there aren't fantastic people who lovingly and devotedly take care of, provide for, and help raise children who are not biologically theirs. Sometimes stepfamilies work perfectly. It did for my friend Charlotte and her sister. After their mother divorced, she began dating a single father who had one young daughter. He had lost his wife to cancer a few years earlier and, like Tom the repairman, was struggling to take care of his young daughter and keep up with his career. After dating for about a year, he and Charlotte's mother decided to take the leap and move in together. The girls bonded immediately as sisters. So much so that the ex-husband agreed to have the "third" daughter come along during his visitation days and weekends. All three daughters are doing better now than before, and even all the adults seem to be living happily ever after. Essentially, the ex-husband adopted this other man's daughter, and the ex-wife's new husband did the same with Charlotte and her sister. As we will see in just a moment, we are not the only species to adopt.

WILD LESSONS

* Similar to the family dynamics of other species, violence against children is more common by stepparents than biological parents.

- Given that unintentional inattentiveness more frequently leads to injury and death when children are being supervised by a stepparent, have a conversation to raise awareness and encourage the nonbiological parent to take steps to become more watchful.
- As a parent, recognize that just as not every bluebird wants to raise a chick that isn't his or hers, neither does every person. Honor and accept when someone isn't willing to step in and fully raise your child.
- Be on the lookout for aggression and poor treatment, especially when you aren't around or when your partner has additional children of his or her own.
- Remember that children are better off with one solid loving parent than a pair that includes a substandard parent, biological or not.

Adoption: Raising Kids Who Aren't Yours

In a very real sense, adoption is much like stepparenting, except that in adoption there are two parents and neither is biologically related to the offspring. On the surface, then, adoption appears to be the ultimate altruistic act. Yet historically, particularly in European and American societies, adoption has served many purposes, the least of which has anything to do with the needs of a child. Orphans often became indentured servants or were otherwise useful to adults (e.g., social climbing, religion, political purposes). For instance, in the early to mid-1800s, orphans were shipped off on trains out West, where families often exploited them as cheap labor.[23] As legislation was put in place to set some standard and supervision over the adoption of children, along with it came secrecy about the identity of a child's biological parents, and even the practice of keeping the knowledge of adoption away from children.

This was the experience of my friend John, a retired Army captain. As he is now sixty, he was adopted in the age of silence, when adoptions were always sealed and the only way to discover who your biological parents really were was through (1) being informed you were adopted and (2) mutual registration, where both parents and children were seeking to reconnect. John's adoption was a bit different from most "traditional" adoptions of that time. His biological parents raised him until he was four, and then, for reasons still unknown to him today, a new family adopted him.

When I asked him if he remembered the day he went to live with his new family, he replied, "Oh yes. On the drive away from home, not understanding what was happening, I screamed for Superman to come and save me. When I got [to my new home], they changed my name from Nick to John and told me they were my new parents." He didn't just lose his parents; he also lost a brother. His new parents were open about the fact that he was adopted, but concealing the truth wasn't really an option, given his advanced age.

Although the reasons for John's adoption remain murky and life with his new parents was troubled at times, he rated the overall transfer to another family a positive experience and a success. We have that advantage as a species: the capacity to look back, reflect, and make a decision about our life experiences. But for scientists, adoption presents a conundrum because it represents a complete investment in someone else's offspring, which, under natural selection, does not benefit one's own reproductive success. However, adoption does increase one's "fitness" in the sense of social acceptance, and it can satisfy a desire to raise a child or fulfill some other need that does not necessarily have anything to do with the needs of a *particular* child. In John's case, his adoptive parents could not have children and wanted a son to carry on the family tradition of attending West Point. Thus, he came into the family with a job to do: go to West Point. Period. Did he benefit? Probably. Was it about helping him specifically? Not really.

When it comes to other species, there are many versions of adoption, and the types we are most familiar with are those where different species seemingly adopt each other—like sows that suckle kittens, or cats that nurse puppies. There is even the infamous YouTube video of a lion "adopting" a baby gazelle. Sometimes this happens, as in the case of the lion, because, as I mentioned in Chapter 3, we and many other species are hardwired to take care of that which bears the hallmarks of infancy.

It is difficult to know the motivations of the individuals involved, and it becomes more complex in cases of adoption within the same species that are more similar to us. Still, one of the explanations is that the foundation and opportunity for adoption is laid by the general attraction to infants. This attraction compels some individuals to act on this to a greater extent by adopting an infant they are not related to. Take the case of an infant Angolan black-and-white colobus monkey. One place you will find these gorgeous primates is deep in the coral rag forests on the coast of Kenya. They normally live in moderately sized social groups containing one or two males, some females, and their offspring.

In early September of 2014, scientists observed something strange happening in what they called the Ufalme group. Malkia, an adult female, had just given birth to Kaskazi, a little boy. One December day a few months later, they noticed Okoa, a four- to five-month-old from another group, on his own, crying and trying to get close to Malkia. Much to their surprise, later that same day, Malkia was seen nursing not just her own son Kaskazi but also Okoa! She continued caring for Okoa alongside her own son, though he did receive less time suckling, being held by her, and in close proximity to her, until Kaskazi, her biological son, died unexpectedly. After his death, Okoa relentlessly insisted on nursing and clinging to Malkia, and she complied, providing him with her full attention. However, the conflict over her own reproductive success and caring for her adopted youngster was evident in that she was actively trying to mate with males while still nursing Okoa, something that would not have happened if her biological son had not died.[24]

The conflict that Malkia experienced is something that can happen in families that choose to adopt and then unexpectedly have a biological child. This of course depends entirely on the parents. Setting the proper example and treating children equally and fairly, whether they are biological or adoptive, will lay the foundation for healthy family dynamics. However, research reveals that sometimes there is a tendency for parents to view problems associated with an adopted child as stemming from some other source outside of the family (e.g., genetics), leading them to make the adoptive child a scapegoat for everything that is wrong with the parents, and even to believe the best solution is to send the child away.[25]

Adoption has been reported in everything from nonhuman primates, rhinoceroses, zebra, various seals, dolphins, emperor penguins, and a suite of birds. But I was curious whether adoption in other species ever bore similarities to John's situation: joining a family with a job to do. I didn't have to look too far before I discovered the Florida scrub jay. Scrub jays are a member of the corvid family that includes magpies, crows, ravens, and nutcrackers. The corvids in general are noted for their intelligence, and scrub jays have been an important model in increasing our understanding of reproductive cooperation, or cooperative breeding.

Normally for scrub jays, helpers are relatives. Typically, one generation of offspring sticks around, helping their parents raise the next generation. The benefit to the parents is that they have help raising more kids and holding down the fort against outsiders, and the benefit to their "adult" children is greater survival and success by holding off on striking out on their own. The tradeoff is not being able to reproduce. It's kind of what life was like in Italy not so long ago. Many of my friends lived with their parents well into their twenties and even thirties to help get a better start in life. Part of that was due to the way housing was limited in Italy, and also to a lack of mortgages. Yes, you used to have to pay for a house in all cash.

Anyway, being a scrub jay is a lot like what being an Italian used to be like. Sometimes, though, unrelated young fledglings, or let's say teenage scrub jays, leave home and join another family. In some sense they are adoptees because they get all the benefits that come with having a family: a territory and secure resources. In exchange for room and board, they become helpers and assist in raising the offspring of their adopted family.[26]

Just like other species, humans across all cultures and all parts of the world adopt nonbiological children for many different reasons, but few maintain the level of secrecy and mystery surrounding many American adoptions. In contrast, when the Māori people of New Zealand adopt (called *Whāngai* by the Māori), the whole community is involved in the decision, and the child is raised knowing all information, including his or her biological parents if they are still alive. This participation of the entire community in the decision to adopt and the collective support of children gives rise to an adage familiar to most of us: "It takes a village." Yet this adage completely contradicts the nuclear family that Americans have come to hold as the gold standard of family and parenting.

Before we move on to the concept of cooperatively raising our children, I want to touch briefly on same-sex partners as another type of family. For a long time, homosexual couples were denied the right to adopt or marry. Oftentimes, to create a family, one partner had a child while the other partner, who may have participated fully in raising the child, had no legal standing as a parent. Fortunately, in recent years this has been changing with the legalization of marriage between same-sex couples and the lifting of restrictions on adoption by these couples. Anyone who still doubts the ability of same-sex couples to successfully parent need only look to the rest of the animal kingdom.

For female Laysan albatrosses in Hawaii, same-sex partnerships are often the only way they can have a family. These large seabirds are principally found on the northwestern islands. Wisdom, a Laysan albatross, has risen to fame worldwide because she is the oldest known bird in the Northern Hemisphere. She was tagged in 1956 and had a chick this year with her longtime male partner. Interestingly, scientists noticed that on the island of Oahu, about 30 percent of the adult pairs were actually females. Even more astounding was the discovery that female pairs were cooperatively raising their young together. And they are the model of a nuclear family, mirroring everything observed in heterosexual albatrosses, with the exception of mating with each other. Another key difference is that they take turns having chicks, as albatrosses can only successfully raise one chick at a time. As a result, each female in a female-female pair has few offspring when paired with a female but more than if she had no partner at all.[27] The reason for this pattern in Laysan albatrosses is a combination of too many females immigrating onto the islands and not enough males. Because they must raise their chicks as a pair to have any hope, female albatrosses team up to do it together.

Women in Tanzania are marrying each other for a different, but related, reason. In the Kurya tribe, there is a tradition called the "house of women," where a woman who is widowed or whose husband has left her is permitted to marry a younger female so that she may retain the family home. The younger female has equal rights to the property and is allowed to take a male partner to produce children. The female couple will then raise the children together. This tradition is seeing resurgence in response to the mistreatment of these women by men in this tribe. By marrying each other, each female increases her resource base, and by cooperatively raising children, their children benefit. I had a standing agreement with one of my closest friends that should we both find ourselves single by the time we were thirty-five, we would marry each other and raise kids together. I never knew I was such an albatross. In the end, she got married and has three beautiful kids. Now our agreement is that should she ever get divorced, we will blend our families together, creating our own "village."

WILD LESSONS

* Adoption is an extreme form of stepparenting and happens in all kinds of animals.
* The reasons for adoption are as varied in humans as other animals and can include getting parenting practice, benefitting from having an extra helper, increasing social standing, or simply feeling the strong pull of attraction to an infant.
* When choosing whether or not to adopt a child, being very clear on the reasons is critical. Failing to do so can lead to problems later, more likely for the adoptee than the adults.
* Children are not scrub jays, yet using a child, biological or not, to fulfill a specific role for the parent may be more common than we like to admit.
* Although open adoptions may be more difficult for the parents, if all parties can agree, it may be better for the child to have as much information as possible.
* Same-sex partnerships are equally valid options for families, from albatrosses to humans.

Does It Really Take a Village?

As we have seen, whether it is via stepparenting, adoption, or forming same-sex partnerships, there are as many ways to create a family as there are to raise children. The Māori practice of communally deciding on when and how an adoption should take place is just one tiny example of what we might collectively refer to as cooperative breeding. But is there a more general sense of community that is important when it comes to raising children?

According to Dr. Sarah Blaffer Hrdy, we are essentially a cooperative breeding species. In her book *Mothers and Others*, she points to the collective and cooperative care of children across most human cultures, contemporary society being a notable exception. For example, 40 to 50 percent of all primate species have what might be loosely considered cooperative care in that mothers willingly allow others to hold or play with their infants. This benefits the mothers primarily by giving them a break and benefits the others with gained experience in caring for an infant. Relying on communal care is adaptive in many other social, group-living species, and how and why others cooperate is as varied as the types of families we have already discussed.

Meerkats rose to fame in the television series *Meerkat Manor* and are considered cooperative breeders. Belonging to the mongoose family, these gregarious small carnivores live in fairly tight-knit social groups, where only the alpha pair breed and everyone else helps take care of their offspring. The type of care that helper meerkats provide may include babysitting, feeding, burrow maintenance, watching out for dangerous predators, comforting distressed pups, and, for females, helping nurse the young. Unlike bees or ants, meerkat helpers don't specialize in providing a particular service,[28] but they are typically under two years old, don't breed, and benefit by staying with their social group. Even though they experience an advantage by being able to stay in their social unit longer, there is also an element of coercion. If a female that is not the alpha female has offspring, then the dominant female will almost certainly kill her babies. This then frees up that female to nurse the young of the alpha female.

Is there a level of cooperative care in other species that is not enforced through oppression, bullying, or punishment? There is when we look at a fascinating, large tropical bird called the green woodhoopoe. Native to forests in Africa, these birds live in groups and feed on insects, and they help each other raise their chicks. Helper green woodhoopoes are there before the eggs are even laid, feeding the mom-to-be. Once the chicks emerge from their eggs, helpers bring food to the mother, who then gives it to the chicks. The big difference between this species and some of the others I have mentioned already is that all adults help at some point, which implies the practice is very reciprocal. Also, helpers aren't only subordinate individuals since there is no dominance structure in this species and, unlike for meerkats, status doesn't matter.[29]

There are a few reasons why this works. First, everyone in the group automatically equally shares the benefits of being in a group, which encourages the development and maintenance of cooperation. Second, groups are small and the cost of helping isn't too high. And third, like humans, parent and helper green woodhoopoes are primed to respond to chicks. Another interesting aspect of this type of cooperative care is that it can be independent of relatedness.

One behavior we have historically had in common with many nonhuman primate species is infant sharing. For dusky leaf monkey infants, being bright orange helps keep them at the center of attention. Allomothering is the norm, and females without infants are busy helping carry around the infants for the first few weeks.

For much of human history, infant sharing was crucial to our survival as a species. As Dr. Hrdy points out, however, we are not born bright orange, but for one tribe with a high incidence of infant sharing, the Beng tribe of West Africa, decorating their newborns helps make them all the more appealing to other women. Beng women face enormous work demands, pretty much taking care of everything, including chopping wood![30] Consequently, the women rely on each other to help take care of each other's infants. This is also true of many hunter-gatherer societies.

I want to address one piece of misinformation about human infants: the myth that they can't bond strongly and equally with a number of adults, or that somehow, by allowing infants to bond with many adults, they will be confused over who their primary caretaker is. The research, and real life examples, do not support this conclusion, and we see that in these types of communities, infants are loved, handled, kissed, cared for, comforted, and fawned over by everyone, relative or not, but still have *more* contact with their mother. That is significant. And this is still true to this day in many societies where the ideal of the nuclear family has not permeated the culture, where some babies may have as many as fourteen caretakers.[31]

Perhaps our tendency to move around more in contemporary society has diluted the help we might have received from family members, but as we see in the cases of the dusky leaf monkey and the green woodhoopoe, support doesn't have to come exclusively from relatives. In a very real sense, modern daycare is the village, but that is very different from a personal network of support. There is no real incentive for an individual not related to you or part of your social circle to provide the highest-quality care.

There is a cognitive dissonance between parents' pleas for help in today's modern families and the insistence on the concept of the nuclear family as the "right" family to have. Children, too, can bond with other children who are not their relatives. There is no one-size-fits-all model. We can learn from other species and cultures where nonrelatives greatly contribute to the raising of the young; and by understanding when and how this can work, we may be able to increase our children's chances of success and bring newly formed families closer than ever.

CHAPTER 10

A Return to Our Roots

W hen I decided to tackle the topic of parenting in humans and animals, I knew it was an ambitious and formidable task. There are a multitude of aspects to parenting, and as I delved deeper, I grew more fascinated with each exciting discovery. The sheer variation in how we and other animals become pregnant and the countless things that have to happen at precisely the right moment to begin the development process that ultimately brings new life into this world is awe-inspiring in itself. Parenting in the animal kingdom is as thought-provoking for its breadth of behaviors as it is for its parallels to the human experience, from male seahorses that experience surprisingly humanlike pregnancy to the cravings and nausea that plague humans and baboons alike. Who would have thought that humans are linked to fruit bats in our common need for the help of a midwife during labor? The list of connections is long, and the examples I've been able to provide in this book only scratch the surface of the common ground that we, as parents, share with other species.

Perhaps the single most unifying element, and what motivated me to write this book, is the dedication and love given by virtually all parents toward the same goal: to create the possibility for offspring to thrive, be healthy, successful, and resilient so they can function in this world as the best version of themselves. When we examine our own parenting behaviors, the question is, *Are we parenting in ways that achieve this goal? And what does it take to raise our children successfully?*

Parenting takes an enormous amount of energy. For humans and for other species, the metabolic, immune, hormonal, and other demands of pregnancy require vast amounts of physical resources—and all of that before an offspring has even been born. Right away there is a conflict between parents and offspring over how much a parent can provide and how much an offspring needs. Fortunately for offspring that depend on parenting, parents are evolutionarily deeply inclined to care for their infants. Babies are cute, exude hormones, deposit their cells into their mother, and unknowingly deploy a suite of measures to ensure their own survival. In the moments immediately after birth, there already exists an intimate exchange between the parent(s) and offspring, solidifying the bond through smell, sight, sound, and touch. Discovering all the ways in which we develop an attachment to the adults who are taking care of us when we are first born opened my eyes to the profound importance of these early bonds and helped me understand why I was so close to my grandmother, Oma.

Through my research and the writing of this book, I also came to a deeper appreciation of the aspects of parenting that so closely connect us with other species—so close, in fact, that we respond to the cries of other species' infants and they, in turn, respond to the cries of ours! This makes our contemporary, widely accepted approach of withdrawing from infants to let them "cry it out" seem contradictory to how we instinctively respond to an infant in distress. In order to grasp how it is that humans are the only animals on the planet who have adopted this strategy, there are many factors to investigate, including culture, technology, and sociology. Then again, perhaps it is a reflection of the deep conflict parents sometimes experience between their own needs and those of their offspring.

So many things change once children arrive on the scene, and few parents seem truly prepared for the extent of the emotional, mental, and physical resources needed to parent well. Here you are with a cute, adorable, completely dependent infant, and your hormones are in flux, you are sleep deprived, and you have little or no time at all for yourself. It is a massive transition that impacts parents in a variety of ways. We have come to recognize that the problem of severe postpartum depression, caused by pronounced hormonal fluctuations that hamper one's ability to care for a newborn effectively, is shared between humans and other species. Understanding that this happens in rats, for example, can hopefully reduce the negative stigma associated with postpartum depression. It isn't the fault of mothers; it is a biological response to a cascade of hormone changes that we truly cannot comprehend even if we can quantify it. Our society isolates new mothers and new families from the communal support they so desperately need, leaving many feeling they have nowhere to turn. The good news is that because we have created this reality, as a society we can come together and move in a new direction, too. Creating a safe, nurturing environment seems to be the key to raising successful offspring. Elephants do this through a community of females that all nurture, teach, and protect their young collectively, while wolf packs rely on extended family to babysit and feed the pups.

We can do the same. During my research, it became clear to me that human parents need support. In other species, like langurs, where new mothers also need support, younger females may step in and help mothers out, carrying their infants, grooming them, and playing with them. In the absence of extended social support in a two-parent household, there can be conflict in dividing up who does what. In many bird species, we witness beautifully orchestrated cooperative care by both parents. In hornbills, the mom "stays home" with the kids, literally closed in until they are about ready to fledge, while the dad does all the work collecting and delivering food. Bushtits solve this dilemma by coordinating their activities more evenly. If we can do the same and take turns, parents (when there are two) will experience less of a drain on their ability to meet some of their own needs, and the children will benefit, too! Studies show that children benefit from having as many adults as possible caring for them.

I think it truly does take a village to raise a child. The question that we should be asking ourselves is how we can transform our modern society to look and act more like a village. Until we do so, we'll continue to needlessly isolate and ostracize parents, to the detriment of both them and their kids. As I researched and wrote this book, I spoke with a wide range of parents, mothers, fathers, single parents, divorced parents, new parents, parents of teenagers, and everything in between. One topic that consistently emerged was breastfeeding— specifically, breastfeeding in public. This was, and remains, mystifying to me, because it's an example of how we've ignored the lessons found in animal behavior all around us and made an aspect of parenting harder than it needs to be. We have the distinction of being the only species for whom breastfeeding is even a question, much less a controversy. One of the more bizarre aspects of the breastfeeding discussion is that we now live in a time when experts in the field of infant and child health are starting to enthusiastically recommend it, because mother's milk has been proven to be the optimal food for our infants. Unfortunately, social norms hinder mothers from successfully feeding their babies in the most natural way possible. I'm going to go ahead and just say it: Breast milk is like magic! It is one of nature's most remarkable substances, and, combined with baby saliva, it is truly a potent concoction! Breastfeeding should be practiced by every mother capable of nursing, and it should be not just tolerated but encouraged by the rest of society.

Breastfeeding is natural, yes, but how to breastfeed requires learning. Animals learn from watching other animals. Humans, until recently, did the same: They learned from observing women in their family circle—mothers, grandmothers, aunts, cousins. When babies had difficulty latching on, there were always lots of "experts" around to help out. So take heart; no one is good at it from the get-go. If you're having problems, you're not alone. Reach out and get some much-needed help.

An interesting aspect of parenting that I had never really considered outside of my studies of wild animals from a survival perspective is how many kids to have. With each child, human parents must face an ever-increasing demand on resources—food, time, money, etc. Animal parents have their own ways of deciding how many offspring to have. And, to my mind, many of their "wild" motives—particularly those that determine whether or not to have offspring, and how many, based on the probability they can adequately care for them—make a lot more sense than how humans approach the same dilemma. There are some exceptions, of course. Armadillos and lynx avoid the issue altogether by always having the same number of children—four and two, respectively—no matter what. For both, it is a matter of physiological constraint. The armadillo reproductive system has evolved to always have quadruplets. For lynx, their environment is unpredictable, and females are limited in how much extra weight they can hold, so the optimal number of cubs to successfully nurse under any condition is two.

Many couples think that two children is the perfect number. More likely than not, they hope for one girl and one boy. Ah, the perfect family. But why? There are a variety of motives behind wanting to have two or more, often disconnected from physiological or ecological factors (e.g., resource availability). Often we have at least two children for the simple reason that we think only children are likelier to be selfish or lonely. This is a misconception. The reality is that any child can be selfish or lonely, with or without a sibling, and having siblings introduces a whole other slew of challenges—including sibling rivalry, which, in some circumstances, can be so severe it leads to death.

Though my research broadened my insight into sibling rivalry, the concept wasn't new to me. I was surprised, however, to find that, for humans, violence among siblings accounts for a large portion of violence in the home. Since we are not tiger sharks, whose siblings eat each other, we should all be taking a good, hard look at these statistics to understand what these numbers are saying about us and about our society as a whole. On a micro level, parents certainly need to take negative interactions between siblings seriously and reduce violence between children by teaching them conflict resolution skills that are rooted in empathy and connection.

Closely related to sibling aggression and violence is parental aggression and punishment, both verbal and physical. As a behavioral scientist, I'm happy to say that, though all too common in human parenting, parental aggression is relatively rare in other species. And as I examined this subject more closely, it was clear that we humans run into trouble for a few reasons. First, we have remarkably unrealistic expectations of our children. Animals seem to have a better sense and intuition of what their offspring are capable of and when. We also have highly unrealistic ideas of how children learn things.

I once saw a mother cheetah bring back a live baby gazelle to her two young, inexperienced cubs. They tried unsuccessfully to kill it, and to a casual observer it might have seemed a miserable effort; the cubs appeared to be playing with it more than seriously trying to kill it. The gazelle escaped a few times, and the mom, who was watching patiently, went off to re-wrangle the poor, terrified baby gazelle, brought it back to the cubs, and let them give it another go. She never got angry or frustrated, nor did she lash out at the two for letting the gazelle run off again and again. I was witnessing the process of proper parental guidance. The cubs had no idea what they were doing, but with patient teaching, they would learn.

I've also observed that discipline and learning go hand in hand. Balancing patient teaching with setting parameters is a critical parental skill. We must be able to make a clear distinction between boundaries and punishment, as they are not the same thing. Dolphin mothers, and other adults, enforce limits on how far away a calf can swim. If a calf goes rogue, parents employ a suite of measures, with physical correction being the absolute *last* resort.

In my search to find definitive answers as to why we humans express aggression and hostility toward our children, I examined factors in the external environment. Through studies, we know that the frequency of these nonadaptive, aggressive, violent behaviors rises in other species when the parent(s) is stressed. In other species, we see what we might classify as physical abuse when a parent is under extreme duress, whether the stress is environmental, physical, or psychological. What does this mean for us? In our modern society, which leaves so many parents with mounting social, economic, and environmental stress, it's imperative to find ways to strike a better, less stressful balance. Failing to do so takes a serious toll on our ability to parent well.

It's a parent's responsibility to teach children the parameters of being part of society and to steer them toward loving, empathetic, and productive adulthood. Sadly, we are accomplishing the opposite. I propose that it is, in part, our inability to connect with our nature, with each other, and with the natural world around us that leads to parenting with aggression and intimidation. We know that children who experience even more mild forms of parental or caretaker hostility experience developmental delays as well as cognitive and psychological difficulties. When we look at animals experiencing the conditions that result in physical mistreatment of their offspring, such as rhesus macaques or Nazca boobies, we see that the pattern becomes intergenerational, even if external conditions improve. This is disconcerting, but it's also extraordinarily similar to what we know happens in humans. If you come from a family where physical, verbal, or psychological abuse occurred, there is a higher likelihood that you will repeat the cycle of violence with your children.

How, then, does one balance discipline with allowing offspring their independence? Children cannot just run amok doing whatever they please, but pushing boundaries is part of natural development. Other animals show us that finding incremental ways to support children's growing independence, while keeping them safe, is a difficult but necessary balance.

It follows that more and more scientists from the behavioral, biological, and social sciences are discovering that empathy is one of the most important characteristics of a healthy adult and a cooperative society. It was a complete revelation for me that we belong to more social groups in our lifetime than any other species. This crucial process of socialization begins in the first moments after birth, and it requires interacting with other family members, with other children, and eventually with members outside our small circle: teachers, coworkers, strangers. Teaching our children the strategies to get along with others in any situation is central to a cooperative society, if that is the society we strive to create. While some of us may be more naturally inclined to be empathetic, parents can foster a greater sense of empathy by modeling it, talking about other people's psychological states and experiences, and providing children with a diversity of social settings. We may lament the high levels of bullying that we observe in children (no to mention adults), but we can take action by nurturing the development of empathy in our children. Danish schools even have a class dedicated to developing empathy where students listen to the challenges their fellow students are experiencing and offer help and support. In this way, as discussed in Chapter 7, we may be lagging behind the humble rat. Rats automatically help each other, even when they must forego a personal reward in order to do so. But the research on rats reveals that biases can emerge when they are not exposed to individuals that look different. Given the social climate we live in, with gender bias, racism, and other discriminatory practices, simply integrating children with others from a variety of backgrounds may go a long way toward changing these negative attitudes.

Another pillar of a cooperative society is the willingness to share. Parents can foster sharing by having children collaborate together, whether with siblings or other children. And though a child may initially resist, trust me, with persistence they'll learn to do it. Even the Little Scrub Island lizard shares when it finds food that is larger than it can consume.

There are so many other lessons to learn from animals and their families, but I will leave you with just one more. Whether you are a single parent, a heterosexual couple, a same-sex couple, or a step-parent with a biological child, an adoptive child, a stepchild, or any combination thereof, all parents have to keep in mind that families come in all shapes, sizes, and colors. There is no single blueprint for a prosperous family, an effective parent, or thriving children. From all my research, I've come to believe that better parenting begins with consciously asking ourselves this question: *What kind of people do we want to raise?* Do we want cooperative, helpful, giving, and sharing people who respect the autonomy of others to make up the majority of our society? Deciding what we want our children to be like and what we want our society to look like is the first step toward making it a reality. The next step is encouraging the behaviors that bring it about.

In this sense, we could decide to be like rhesus macaques, despotic and selfish. Or we could decide to be more like the Tonkean black macaques, who live in a society that has less aggression, resolves conflicts easily and with less violence, and has more peaceful interactions overall. Even if we look exclusively at humans, there are principles of egalitarian societies we can emulate—specifically, for example, the hunter-gatherers of the Congo Basin. Their approach as a community is one where cooperation, sharing, and respect is central in their dealings with one another. How do they achieve this? Partly through enlightened parenting.

First, there is always physical and emotional closeness. Adults are always touching each other, and infants are held up to 91 percent of the day. Four-year-olds are held 44 percent of the time. All of that touching is very similar to how baboons forge strong connections with each other. They get their touching through grooming, which has the added benefit of reducing ticks. Of course, parents can't be expected to hold their children half the day, but frequently touching them, holding them, cuddling with them, and letting them observe physical affection between adults is definitely doable.

It reduces conflict among both adults and children, whether or not they are biologically related.

Another key feature of healthy parenting is granting children autonomy over their daily life. For us, that may simply mean not micromanaging every single aspect of our children's day. If I could, I'd send a message to the mom of my younger self and say: "Stop worrying about what other people think and let her wear that crazy outfit she picked out!" Obviously, we need some limits in place to keep our kids safe, but letting them explore who they are and what they like is an important part of respecting them as individuals.

And finally, play! Children need to play. In hunter-gatherer societies, play is still ranked high on the list of things that are of significant importance. Kids are allowed to play and lie around half the day. In today's modern society, our kids get thirty minutes of play and six hours of homework! We've even structured play to the point that it is a second job for many kids, after their homework. In contrast, where much of our play is centered on competition and winning, children in these hunter-gatherer societies engage in a rich and frequent free-style social play that promotes cooperation, sharing, and nonaggressive interactions with others. Play is crucial for many animals. Those little degus romp around, tumbling together as part of their own natural, physical, and social development. Let's bring back unstructured play for our children.

This is a turning point in our society. As our knowledge expands, we have, now more than ever, all the information at our fingertips to become a better, smarter, more connected society. It's time to choose. I know what my choice is. My choice is to take part in creating a more cooperative and empathetic culture. I am optimistic that such a culture is within reach, and the path starts with parenting. For successful parenting to become prevalent, we must create an environment that supports that process. What can help? Connection, to each other and to the natural world of which we are an intimate part. *All* parents need the support of other adults. More than ever, I am convinced that supporting each other—in a community or village, town or city, or even in just a close-knit personal network—is necessary. Let's create our own villages, and let's not forget to stop, admire, and appreciate our biological connection to all the other animal parents!

NOTES

CHAPTER 1

1. Fairbanks, L. A., and M. T. McGuire, "Maternal Condition and the Quality of Maternal Care in Vervet Monkeys," *Behaviour* 132, no. 9 (1995): 733–54.

2. Trivers, Robert L., "Parent-Offspring Conflict," *American Zoologist* 14, no. 1 (1974): 249–64.

3. Atkinson, S. N., and M. A. Ramsay, "The Effects of Prolonged Fasting of the Body Composition and Reproductive Success of Female Polar Bears (*Ursus maritimus*)," *Functional Ecology* 9 (1995): 559–67.

4. Ligon, Russell A., and Geoffrey E. Hill, "Feeding Decisions of Eastern Bluebirds Are Situationally Influenced by Fledgling Plumage Color," *Behavioral Ecology* 21, no. 3 (2010): 456–64.

5. Leonard, M. L., A. G. Horn, and S. F. Eden, "Parent-Offspring Aggression in Moorhens," *Behavioral Ecology and Sociobiology* 23, no. 4 (1988): 265–70.

CHAPTER 2

1. Tyler, Michael J., F. Seamark, and R. Kelly, "Inhibition of Gastric Acid Secretion in the Gastric Brooding," *Science* 220 (1983): 609–10.

2. Bouchie, Lynette, et al., "Are Cape Ground Squirrels (*Xerus inauris*) Induced or Spontaneous Ovulators?" *Journal of Mammalogy* 87, no. 1 (2006): 60–66.

3. Tomlinson, M. J., et al., "The Removal of Morphologically Abnormal Sperm Forms by Phagocytes: A Positive Role for Seminal Leukocytes?" *Human Reproduction* 7, no. 4 (1992): 517–22.

4. Whittington, Camilla M., et al., "Seahorse Brood Pouch Transcriptome Reveals Common Genes Associated with Vertebrate Pregnancy," *Molecular Biology and Evolution* 32, no. 12 (2015): 3114–31.

5. Ziegler, Toni E., et al., "Pregnancy Weight Gain: Marmoset and Tamarin Dads Show It Too," *Biology Letters* 2, no. 2 (2006): 181–83.

6. de Souza Leite, Melina, et al., "Activity Patterns of the Water Opossum *Chironectes minimus* in Atlantic Forest Rivers of South-Eastern Brazil," *Journal of Tropical Ecology* 29, no. 03 (2013): 261–64.

7. Jukic, A. M., et al., "Length of Human Pregnancy and Contributors to Its Natural Variation," *Human Reproduction* 28, no. 10 (2013): 2848–55.

8. Tanaka, S., et al., "The Reproductive Biology of the Frilled Shark, *Chlamydoselachus anguineus*, from Suruga Bay, Japan," *Japanese Journal of Ichthyology (Japan)* 37, no. 03 (1990): 273–91.

9. Ptak, Grazyna E., Jacek A. Modlinski, and Pasqualino Loi, "Embryonic Diapause in Humans: Time to Consider?" *Reproductive Biology and Endocrinology* 11, no. 1 (2013): 1.

10. Grinsted, Jørgen, and Birthe Avery. "A Sporadic Case of Delayed Implantation After In-Vitro Fertilization in the Human?" *Human Reproduction* 11, no. 3 (1996): 651–54.

11. Schuett, G. W., et al., "Unlike Most Vipers, Female Rattlesnakes (*Crotalus atrox*) Continue to Hunt and Feed Throughout Pregnancy," *Journal of Zoology* 289, no. 2 (2013): 101–10.

12. Morisaki, Naho, et al., "Declines in Birth Weight and Fetal Growth Independent of Gestational Length," *Obstetrics and Gynecology* 121, no. 1 (2013): 51.

13. Asbee, Shelly M., et al., "Preventing Excessive Weight Gain During Pregnancy Through Dietary and Lifestyle Counseling: A Randomized Controlled Trial," *Obstetrics and Gynecology* 113, no. 2, Part 1 (2009): 305–12.

14. Centers for Disease Control and Prevention, "Gestational Weight Gain—United States, 2012 and 2013," *Morbidity and Mortality Weekly Report* 64, no. 43 (November 6, 2015): 1215–20.

15. Whitaker, Robert C., "Predicting Preschooler Obesity at Birth: The Role of Maternal Obesity in Early Pregnancy," *Pediatrics* 114, no. 1 (2004): e29–e36.

16. Colodro-Conde, et al., "Nausea and Vomiting During Pregnancy Is Highly Heritable," *Behavior Genetics* (2016): 1–11.

17. Profet, Margie, "The Evolution of Pregnancy Sickness as Protection to the Embryo Against Pleistocene Teratogens," *Evolutionary Theory* 8 (1988): 177–90.

18. Mor, Gil, and Ingrid Cardenas, "Review Article: The Immune System in Pregnancy: A Unique Complexity," *American Journal of Reproductive Immunology* 63, no. 6 (2010): 425–33.

19. Kourtis, Athena P., Jennifer S. Read, and Denise J. Jamieson, "Pregnancy and Infection," *New England Journal of Medicine* 370, no. 23 (2014): 2211–18.

20. Cousins, Don, and Michael A. Huffman, "Medicinal Properties in the Diet of Gorillas: An Ethnopharmacological Evaluation," *African Study Monographs* 23, no. 2 (2002): 65–89.

21. Czaja, John A., "Food Rejection by Female Rhesus Monkeys During the Menstrual Cycle and Early Pregnancy," *Physiology and Behavior* 14, no. 5 (1975): 579–87.

22. Young, Sera L., "Pica in Pregnancy: New Ideas About an Old Condition," *Annual Review of Nutrition* 30 (2010): 403–22.

23. Fawcett, Emily J., Jonathan M. Fawcett, and Dwight Mazmanian, "A Meta-Analysis of the Worldwide Prevalence of Pica During Pregnancy and the Postpartum Period," *International Journal of Gynecology and Obstetrics* 133, no. 3 (2016): 277–83.

24. Pebsworth, Paula A., Massimo Bardi, and Michael A. Huffman, "Geophagy in Chacma Baboons: Patterns of Soil Consumption by Age, Class, Sex, and Reproductive State," *American Journal of Primatology* 74, no. 1 (2012): 48–57.

25. Teyssier, Jérémie, et al., "Photonic Crystals Cause Active Colour Change in Chameleons," *Nature Communications* 6 (2015).

26. Anderson, Marla V., and M. D. Rutherford, "Evidence of a Nesting Psychology During Human Pregnancy," *Evolution and Human Behavior* 34, no. 6 (2013): 390–97.

27. Northrop, Lesley E., and Nancy Czekala, "Reproduction of the Red Panda," in *Red Panda: Biology and Conservation of the First Panda*, ed. Angela R. Glatston (Burlington, MA: Elsevier, 2010), 125.

28. Harcourt, Caroline, "*Galago zanzibaricus*: Birth Seasonally, Litter Size and Perinatal Behaviour of Females," *Journal of Zoology* 210, no. 3 (1986): 451–57.

29. Estes, Richard D., and Runhild K. Estes, "The Birth and Survival of Wildebeest Calves," *Zeitschrift für Tierpsychologie* 50, no. 1 (1979): 45–95.

30. Ji, R., et al., "Monophyletic Origin of Domestic Bactrian Camel (*Camelus bactrianus*) and Its Evolutionary Relationship with the Extant Wild Camel (*Camelus bactrianus ferus*)," *Animal Genetics* 40, no. 4 (2009): 377–82.

31. Lenssen-Erz, Tilman, "Adaptation or Aesthetic Alleviation: Which Kind of Evolution Do We See in Saharan Herder Rock Art of Northeast Chad?" *Cambridge Archaeological Journal* 22, no. 01 (2012): 89–114.

32. Niasari-Naslaji, Amir, "An Update on Bactrian Camel Reproduction," *Journal of Camel Practice and Research* 15 (2008): 1–6.

33. Hartwig, Walter Carl, "Effect of Life History on the Squirrel Monkey (Platyrrhini, Saimiri) Cranium," *American Journal of Physical Anthropology* 97, no. 4 (1995): 435–49.

34. Trevathan, Wenda, "Primate Pelvic Anatomy and Implications for Birth," *Philosophical Transactions of the Royal Society B: Biological Sciences* 370, no. 1663 (2015): 20140065.

35. Ibid.

36. Jones, Jennifer S., and Katherine E. Wynne-Edwards, "Paternal Hamsters Mechanically Assist the Delivery, Consume Amniotic Fluid and Placenta, Remove Fetal Membranes, and Provide Parental Care During the Birth Process," *Hormones and Behavior* 37, no. 2 (2000): 116–25.

37. Kunz, T. H., et al., "Allomaternal Care: Helper?Assisted Birth in the Rodrigues Fruit Bat, *Pteropus rodricensis* (Chiroptera: Pteropodidae)," *Journal of Zoology* 232, no. 4 (1994): 691–700.

38. Centers for Disease Control and Prevention, "Pregnancy Mortality Surveillance System," cdc.gov/reproductivehealth/maternalinfanthealth/pmss.html, accessed January 5, 2017.

CHAPTER 3

1. Lorenz, Konrad, "Die angeborenen Formen möglicher Erfahrung," *Zeitschrift für Tierpsychologie* 5 (1943): 94–125; Vicedo, Marga, "The Father of Ethology and the Foster Mother of Ducks: Konrad Lorenz as Expert on Motherhood," *Isis* 100, no. 2 (2009): 263–91.

2. Nittono, Hiroshi, et al., "The Power of Kawaii: Viewing Cute Images Promotes a Careful Behavior and Narrows Attentional Focus," *PloS one* 7, no. 9 (2012): e46362; Sherman, Gary D., Jonathan Haidt, and James A. Coan, "Viewing Cute Images Increases Behavioral Carefulness," *Emotion* 9, no. 2 (2009): 282.

3. Ibid.

4. Casey, Rita J., and Jean M. Ritter, "How Infant Appearance Informs: Child Care Providers' Responses to Babies Varying in Appearance of Age and Attractiveness," *Journal of Applied Developmental Psychology* 17, no. 4 (1996): 495–518.

5. Ritter, Jean M., Rita J. Casey, and Judith H. Langlois, "Adults' Responses to Infants Varying in Appearance of Age and Attractiveness," *Child Development* 62, no. 1 (1991): 68–82.

6. Soler, Manuel, Tomás Pérez-Contreras, and Liesbeth Neve. "Great Spotted Cuckoos Frequently Lay Their Eggs While Their Magpie Host Is Incubating," *Ethology* 120, no. 10 (2014): 965–72.

7. Stoddard, Mary Caswell, and Martin Stevens, "Pattern Mimicry of Host Eggs by the Common Cuckoo, as Seen Through a Bird's Eye," *Proceedings of the Royal Society of London B: Biological Sciences* (2010): rspb20092018.

8. Soler, Manuel, et al., "Preferential Allocation of Food by Magpies *Pica pica* to Great Spotted Cuckoo Clamator Glandarius Chicks," *Behavioral Ecology and Sociobiology* 37, no. 1 (1995): 7–13.

9. Alvergne, Alexandra, Charlotte Faurie, and Michel Raymond, "Are Parents' Perceptions of Offspring Facial Resemblance Consistent with Actual Resemblance? Effects on Parental Investment," *Evolution and Human Behavior* 31, no. 1 (2010): 7–15.

10. DeBruine, Lisa M., "Facial Resemblance Enhances Trust," *Proceedings of the Royal Society of London B: Biological Sciences* 269, no. 1498 (2002): 1307–12.

11. Miller, Don E., and John T. Emlen, "Individual Chick Recognition and Family Integrity in the Ring-Billed Gull," *Behaviour* 52, no. 1 (1974): 124–43.

12. Porter, Richard H., Jennifer M. Cernoch, and Rene D. Balogh. "Recognition of Neonates by Facial-Visual Characteristics," *Pediatrics* 74, no. 4 (1984): 501–4.

13. Cleveland, Cutler J., et al., "Economic Value of the Pest Control Service Provided by Brazilian Free-Tailed Bats in South-Central Texas," *Frontiers in Ecology and the Environment* 4, no. 5 (2006): 238–43.

14. Gelfand, Deborah L., and Gary F. McCracken. "Individual Variation in the Isolation Calls of Mexican Free-Tailed Bat Pups (*Tadarida brasiliensis mexicana*)," *Animal Behaviour* 34, no. 4 (1986): 1078–86; Balcombe, Jonathan P., "Vocal Recognition of Pups by Mother Mexican Free-Tailed Bats, *Tadarida brasiliensis mexicana*," *Animal Behaviour* 39, no. 5 (1990): 960–66.

15. Dobson, F. Stephen, and Pierre Jouventin, "How Mothers Find Their Pups in a Colony of Antarctic Fur Seals," *Behavioural Processes* 61, no. 1 (2003): 77–85; Charrier, Isabelle, Nicolas Mathevon, and Pierre Jouventin, "Fur Seal Mothers Memorize Subsequent Versions of Developing Pups' Calls: Adaptation to Long-Term Recognition or Evolutionary By-Product?" *Biological Journal of the Linnean Society* 80, no. 2 (2003): 305–12.

16. Formby, David, "Maternal Recognition of Infant's Cry," *Developmental Medicine and Child Neurology* 9, no. 3 (1967): 293–98.

17. Insley, Stephen J., Rosana Paredes, and Ian L. Jones, "Sex Differences in Razorbill *Alca torda* Parent—Offspring Vocal Recognition," *Journal of Experimental Biology* 206, no. 1 (2003): 25–31.

18. Gustafsson, Erik, et al., "Fathers Are Just as Good as Mothers at Recognizing the Cries of Their Baby," *Nature Communications* 4 (2013): 1698.

19. Fripp, Deborah, and Peter Tyack, "Postpartum Whistle Production in Bottlenose Dolphins," *Marine Mammal Science* 24, no. 3 (2008): 479–502.

20. DeCasper, A., and W. Fifer, "Of Human Bonding: Newborns Prefer Their Mothers' Voices," in *Readings on the Development of Children*, eds. M. Gauvain and M. Cole (New York: Worth Publishers: 2004), 56.

21. Moon, Christine, Robin Panneton Cooper, and William P. Fifer, "Two-Day-Olds Prefer Their Native Language," *Infant Behavior and Development* 16, no. 4 (1993): 495–500.

22. Lundström, J. N., et al., "Maternal Status Regulates Cortical Responses to the Body Odor of Newborns," *Frontiers in Psychology* 4 (2013): 597.

23. Corona, R., and F. Lévy, "Chemical Olfactory Signals and Parenthood in Mammals," *Hormones and Behavior* 68 (2015): 77–90.

24. Lévy, F., M. Keller, and P. Poindron, "Olfactory Regulation of Maternal Behavior in Mammals," *Hormones and Behavior* 46, no. 3 (2004): 284–302.

25. Varendi, Heili, et al., "Soothing Effect of Amniotic Fluid Smell in Newborn Infants," *Early Human Development* 51, no. 1 (1998): 47–55.

26. Corona and Lévy, "Chemical Olfactory Signals and Parenthood in Mammals."

27. Bull, C. M., et al., "Recognition of Offspring by Females of the Australian Skink, *Tiliqua rugosa*," *Journal of Herpetology* 28, no. 1 (1994): 117–20.

28. Contreras, Carlos M., et al., "Amniotic Fluid Elicits Appetitive Responses in Human Newborns: Fatty Acids and Appetitive Responses," *Developmental Psychobiology* 55, no. 3 (2013): 221–31.

29. Dageville, C., et al., "Il faut protéger la rencontre de la mère et de son nouveau-né autour de la naissance," *Archives de Pédiatrie* 18, no. 9 (2011): 994–1000.

30. Kaitz, Marsha, et al., "Infant Recognition by Tactile Cues," *Infant Behavior and Development* 16, no. 3 (1993): 333–41.

31. Bader, Alan P., and Roger D. Phillips, "Fathers' Proficiency at Recognizing Their Newborns by Tactile Cues," *Infant Behavior and Development* 22, no. 3 (1999): 405–9.

32. Bystrova, K., et al., "Skin-to-Skin Contact May Reduce Negative Consequences of 'The Stress Of Being Born': A Study on Temperature in Newborn Infants, Subjected to Different Ward Routines in St. Petersburg," *Acta Paediatrica* 92, no. 3 (2003): 320–26.

33. Erlandsson, Kerstin, et al., "Skin-to-Skin Care with the Father After Cesarean Birth and Its Effect on Newborn Crying and Prefeeding Behavior," *Birth* 34, no. 2 (2007): 105–14.

34. Haff, Tonya M., and Robert D. Magrath, "Calling at a Cost: Elevated Nestling Calling Attracts Predators to Active Nests," *Biology Letters* 7, no. 4 (2011): 493–95.

35. Newman, John D., "Neural Circuits Underlying Crying and Cry Responding in Mammals," *Behavioural Brain Research* 182, no. 2 (2007): 155–65.

36. Kushnick, Geoff, "Parental Supply and Offspring Demand Amongst Karo Batak Mothers and Children," *Journal of Biosocial Science* 41, no. 02 (2009): 183–93.

CHAPTER 4

1. Jaimez, N. A., et al., "Urinary Cortisol Levels of Gray-Cheeked Mangabeys Are Higher in Disturbed Compared to Undisturbed Forest Areas in Kibale National Park, Uganda," *Animal Conservation* 15, no. 3 (2012): 242–47.

2. Mulder, Eduard JH, et al., "Prenatal Maternal Stress: Effects on Pregnancy and the (Unborn) Child," *Early Human Development* 70, no. 1 (2002): 3–14.

3. Power, Michael L., and Jay Schulkin, "Functions of Corticotropin-Releasing Hormone in Anthropoid Primates: From Brain to Placenta," *American Journal of Human Biology* 18, no. 4 (2006): 431–47.

4. Bardi, Massimo, et al., "The Role of the Endocrine System in Baboon Maternal Behavior," *Biological Psychiatry* 55, no. 7 (2004): 724–32.

5. Lonstein, Joseph S., Frédéric Lévy, and Alison S. Fleming. "Common and Divergent Psychobiological Mechanisms Underlying Maternal Behaviors in Non-Human and Human Mammals," *Hormones and Behavior* 73 (2015): 156–85.

6. Berg, S. J., and K. E. Wynne-Edwards, "Salivary Hormone Concentrations in Mothers and Fathers Becoming Parents Are Not Correlated," *Hormones and Behavior* 42, no. 4 (2002): 424–36.

7. Miller, David A., Carol M. Vleck, and David L. Otis, "Individual Variation in Baseline and Stress-Induced Corticosterone and Prolactin Levels Predicts Parental Effort by Nesting Mourning Doves," *Hormones and Behavior* 56, no. 4 (2009): 457–64.

8. Van Roo, Brandi L., Ellen D. Ketterson, and Peter J. Sharp, "Testosterone and Prolactin in Two Songbirds That Differ in Paternal Care: The Blue-Headed Vireo and the Red-Eyed Vireo," *Hormones and Behavior* 44, no. 5 (2003): 435–41.

9. Schradin, Carsten, et al., "Prolactin and Paternal Care: Comparison of Three Species of Monogamous New World Monkeys (*Callicebus cupreus, Callithrix jacchus,* and *Callimico goeldii*)," *Journal of Comparative Psychology* 117, no. 2 (2003): 166.

10. Storey, Anne E., et al., "Hormonal Correlates of Paternal Responsiveness in New and Expectant Fathers," *Evolution and Human Behavior* 21, no. 2 (2000): 79–95.

11. Rilling, James K., "The Neural and Hormonal Bases of Human Parental Care," *Neuropsychologia* 51, no. 4 (2013): 731–47.

12. Kunz, Thomas H., and David J. Hosken, "Male Lactation: Why, Why Not and Is It Care?" *Trends in Ecology and Evolution* 24, no. 2 (2009): 80–85.

13. Feldman, Ruth, et al., "Evidence for a Neuroendocrinological Foundation of Human Affiliation: Plasma Oxytocin Levels Across Pregnancy and the Postpartum Period Predict Mother-Infant Bonding," *Psychological Science* 18, no. 11 (2007): 965–70.

14. Bakermans-Kranenburg, Marian J., and Marinus H. van IJzendoorn, "Oxytocin Receptor (*OXTR*) and Serotonin Transporter (*5-HTT*) Genes Associated with Observed Parenting," *Social Cognitive and Affective Neuroscience* 3, no. 2 (2008): 128–34.

15. Saito, Atsuko, and Katsuki Nakamura, "Oxytocin Changes Primate Paternal Tolerance to Offspring in Food Transfer," *Journal of Comparative Physiology A* 197, no. 4 (2011): 329–37.

16. Gordon, Ilanit, et al., "Prolactin, Oxytocin, and the Development of Paternal Behavior Across the First Six Months of Fatherhood," *Hormones and Behavior* 58, no. 3 (2010): 513–18; Rilling, James K., "The Neural and Hormonal Bases of Human Parental Care," *Neuropsychologia* 51, no. 4 (2013): 731–47.

17. O'Hara, Michael W., and Jennifer E. McCabe, "Postpartum Depression: Current Status and Future Directions," *Annual Review of Clinical Psychology* 9 (2013): 379–407.

18. Stoffel, Erin C., and Rebecca M. Craft, "Ovarian Hormone Withdrawal-Induced 'Depression' in Female Rats," *Physiology and Behavior* 83, no. 3 (2004): 505–13.

19. Pereira, Mariana, and Annabel Ferreira, "Affective, Cognitive, and Motivational Processes of Maternal Care," in *Perinatal Programming of Neurodevelopment*, ed. Marta C. Antonelli (New York: Springer, 2015), 199–217.

20. Winnicott, Donald W. "The Capacity to Be Alone," *The International Journal of Psycho-Analysis* 39 (1958): 416.

21. Mayes, Linda C., James E. Swain, and James F. Leckman, "Parental Attachment Systems: Neural Circuits, Genes, and Experiential Contributions to Parental Engagement," *Clinical Neuroscience Research* 4, no. 5 (2005): 301–13.

22. Steyaert, S. M. J. G., et al., "Human Shields Mediate Sexual Conflict in a Top Predator," *Proceedings of the Royal Society of London B: Biological Sciences* 283, no. 1833 (2016): 20160906.

23. Boddy, Amy M., et al., "Fetal Microchimerism and Maternal Health: A Review and Evolutionary Analysis of Cooperation and Conflict Beyond the Womb," *Bioessays* 37, no. 10 (2015): 1106–18.

24. Burkett, J. P., et al., "Oxytocin-Dependent Consolation Behavior in Rodents," *Science* 351, no. 6271 (2016): 375–78.

25. Ruscio, Michael G., et al., "Pup Exposure Elicits Hippocampal Cell Proliferation in the Prairie Vole," *Behavioural Brain Research* 187, no. 1 (2008): 9–16.

26. Anderson, Marla V., and Mel D. Rutherford, "Cognitive Reorganization During Pregnancy and the Postpartum Period: An Evolutionary Perspective," *Evolutionary Psychology* 10, no. 4 (2012): 659–87.

27. Leuner, Benedetta, Erica R. Glasper, and Elizabeth Gould, "Parenting and Plasticity," *Trends in Neurosciences* 33, no. 10 (2010): 465–73; Rolls, A. H. Schori, A. London, and M. Schwartz, "Decrease in Hippocampal Neurogenesis During Pregnancy: A Link to Immunity," *Molecular Psychiatry* 13 (2008): 468–69.

28. Kozorovitskiy, Yevegenia, et al., "Fatherhood Affects Dendritic Spines and Vasopressin V1a Receptors in the Primate Prefrontal Cortex," *Nature Neuroscience* 9 (2006): 1094–95.

29. Stanford, Craig B. "Costs and Benefits of Allomothering in Wild Capped Langurs (*Presbytis pileata*)," *Behavioral Ecology and Sociobiology* 30, no. 01 (1992): 29–34.

30. Bebbington, Kat, and Ben J. Hatchwell, "Coordinated Parental Provisioning Is Related to Feeding Rate and Reproductive Success in a Songbird," *Behavioral Ecology* 27, no. 2 (2016): 652–59.

31. Grand, Robert, "A Collective Case Study of Expectant Father Fears" (PhD dissertation, Liberty University, 2015).

32. Estes and Estes, "The Birth and Survival of Wildebeest Calves."

CHAPTER 5

1. Salomon, Mor, et al., "Maternal Nutrition Affects Offspring Performance Via Maternal Care in a Subsocial Spider," *Behavioral Ecology and Sociobiology* 65, no. 6 (2011): 1191–202.

2. Toyama, Masatoshi, "Adaptive Advantages of Matriphagy in the Foliage Spider, *Chiracanthium japonicum (*Araneae: Clubionidae*)," *Journal of Ethology* 19, no. 2 (2001): 69–74.

3. Kupfer, Alexander, et al., "Parental Investment by Skin Feeding in a Caecilian Amphibian," *Nature* 440, no. 7086 (2006): 926–29.

4. Jenness, Robert, "Biosynthesis and Composition of Milk," *Journal of Investigative Dermatology* 63, no. 1 (1974): 109–18.

5. *Mammals Suck . . . Milk!*, mammalssuck.blogspot.com.

6. Al-Shehri, Saad S., et al., "Breastmilk-Saliva Interactions Boost Innate Immunity by Regulating the Oral Microbiome in Early Infancy," *PloS one* 10, no. 9 (2015): e0135047.

7. Abello, M. T., and M. Colell, "Analysis of Factors That Affect Maternal Behaviour and Breeding Success in Great Apes in Captivity," *International Zoo Yearbook* 40, no. 1 (2006): 323–40.

8. Volk, Anthony A. "Human Breastfeeding Is Not Automatic: Why That's So and What It Means for Human Evolution," *Journal of Social, Evolutionary, and Cultural Psychology* 3, no. 4 (2009): 305.

9. Brown, Amy, "Breast Is Best, But Not in My Back-Yard," *Trends in Molecular Medicine* 21, no. 2 (2015): 57–59.

10. Lydersen, C., K. M. Kovacs, and M. O. Hammill, "Energetics During Nursing and Early Postweaning Fasting in Hooded Seal (*Cystophora cristata*) Pups from the Gulf of St Lawrence, Canada," *Journal of Comparative Physiology B* 167, no. 2 (1997): 81–88.

11. van Noordwijk, et al., "Multi-Year Lactation and its Consequences in Bornean Orangutans (*Pongo pygmaeus wurmbii*)," *Behavioral Ecology and Sociobiology* 67, no. 5 (2013): 805–14.

12. Harvey, Paul H., and Timothy H. Clutton-Brock, "Life History Variation in Primates," *Evolution* 39, no. 3 (1985): 559–81.

13. "Benefits of Babywearing," askdrsears.com/topics/health-concerns/fussy-baby/baby-wearing/benefits-babywearing.

14. Cortez, Michelle, et al., "Development of an Altricial Mammal at Sea: I. Activity Budgets of Female Sea Otters and Their Pups in Simpson Bay, Alaska," *Journal of Experimental Marine Biology and Ecology* 481 (2016): 71–80.

15. Eason, R. R., "Maternal Care as Exhibited by Wolf Spiders (Lycosids)," *Proceedings of the Arkansas Academy of Science* 43 (1964): 13–19.

16. Dionísio, Jadiane, et al., "Palmar Grasp Behavior in Full-Term Newborns in the First 72 Hours of Life," *Physiology and Behavior* 139 (2015): 21–25.

17. Meaney, Michael J., Elizabeth Lozos, and Jane Stewart, "Infant Carrying by Nulliparous Female Vervet Monkeys (*Cercopithecus aethiops*)," *Journal of Comparative Psychology* 104, no. 4 (1990): 377.

18. Gubernick, David J., and Jeffrey R. Alberts, "The Biparental Care System of the California Mouse, *Peromyscus californicus*," *Journal of Comparative Psychology* 101, no. 2 (1987): 169.

19. Keller, Meret A., and Wendy A. Goldberg, "Co-Sleeping: Help or Hindrance for Young Children's Independence?" *Infant and Child Development* 13, no. 5 (2004): 369–88.

20. Ibid.

21. McKenna, James J., and Thomas McDade, "Why Babies Should Never Sleep Alone: A Review of the Co-Sleeping Controversy in Relation to SIDS, Bedsharing and Breast Feeding," *Paediatric Respiratory Reviews* 6, no. 2 (2005): 134–52.

22. Davies, D. P., "Cot Death in Hong Kong: A Rare Problem?" *The Lancet* 326, no. 8468 (1985): 1346–49.

23. Balarajan, R., V. Soni Raleigh, and B. Botting, "Sudden Infant Death Syndrome and Postneonatal Mortality in Immigrants in England and Wales," *BMJ* 298, no. 6675 (1989): 716–20.

24. McKenna, James J., "An Anthropological Perspective on the Sudden Infant Death Syndrome (SIDS): The Role of Parental Breathing Cues and Speech Breathing Adaptations," *Medical Anthropology* 10, no. 1 (1986): 9–53.

25. McKenna and McDade, "Why Babies Should Never Sleep Alone: A Review of the Co-Sleeping Controversy in Relation to SIDS, Bedsharing and Breast Feeding."

26. Pew Research Center, "Parental Time Use," pewresearch.org/data-trend/society-and-demographics/parental-time-use, accessed January 11, 2017.

27. Funston, P. J., M. G. L. Mills, and H. C. Biggs, "Factors Affecting the Hunting Success of Male and Female Lions in the Kruger National Park," *Journal of Zoology* 253, no. 4 (2001): 419–31.

28. Włodarczyk, Radosław, and Piotr Minias, "Division of Parental Duties Confirms a Need for Bi-Parental Care in a Precocial Bird, the Mute Swan Cygnus olor," *Animal Biology* 65, no. 2 (2015): 163–76.

29. Awata, Satoshi, and Masanori Kohda, "Parental Roles and the Amount of care in a Bi-Parental Substrate Brooding Cichlid: The Effect of Size Differences Within Pairs," *Behaviour* 141, no. 9 (2004): 1135–49.

30. Szabó, Nóra, et al., "Understanding Human Biparental Care: Does Partner Presence Matter?" *Early Child Development and Care* 181, no. 5 (2011): 639–47.

31. Kinnaird, Margaret F., and Timothy G. O'Brien, "Breeding Ecology of the Sulawesi Red-Knobbed Hornbill *Aceros cassidix*," *Ibis* 141, no. 1 (1999): 60–69.

32. Markman, Shai, Yoram Yom-Tov, and Jonathan Wright, "Male Parental Care in the Orange-Tufted Sunbird: Behavioural Adjustments in Provisioning and Nest Guarding Effort," *Animal Behaviour* 50, no. 3 (1995): 655–69.

33. Rotkirch, Anna, and Kristiina Janhunen, "Maternal Guilt," *Evolutionary Psychology*, 8, no. 1 (2010): 90–106.

34. Erikstad, Kjell Einar, et al., "Adjustment of Parental Effort in the Puffin: The Roles of Adult Body Condition and Chick Size," *Behavioral Ecology and Sociobiology* 40, no. 2 (1997): 95–100.

35. Rotkirch and Janhunen, "Maternal Guilt."

36. Franks, Nigel R., et al., "Speed Versus Accuracy in Collective Decision Making," *Proceedings of the Royal Society of London B: Biological Sciences* 270, no. 1532 (2003): 2457–63.

37. Bouskila, Amos, and Daniel T. Blumstein, "Rules of Thumb for Predation Hazard Assessment: Predictions from a Dynamic Model," *American Naturalist* (1992): 161–76.

CHAPTER 6

1. Hoffman, Kristi L., K. Jill Kiecolt, and John N. Edwards, "Physical Violence Between Siblings: A Theoretical and Empirical Analysis," *Journal of Family Issues* 26, no. 8 (2005): 1103–30.

2. Loughry, W. J., et al., "Polyembryony in Armadillos: An Unusual Feature of the Female Nine-Banded Armadillo's Reproductive Tract May Explain Why Her Litters Consist of Four Genetically Identical Offspring," *American Scientist* 86, no. 3 (1998): 274–79.

3. Jönsson, Erik G., et al., "Further Studies on a Male Monozygotic Triplet with Schizophrenia: Cytogenetical and Neurobiological Assessments in the Patients and Their Parents," *European Archives of Psychiatry and Clinical Neuroscience* 247, no. 5 (1997): 239–47.

4. Sekar, Aswin, et al., "Schizophrenia Risk from Complex Variation of Complement Component 4," *Nature* 530, no. 7589 (2016): 177–83.

5. Gaillard, Jean-Michel, et al., "One Size Fits All: Eurasian Lynx Females Share a Common Optimal Litter Size," *Journal of Animal Ecology* 83, no. 1 (2014): 107–15.

6. Lawson, David W., and Ruth Mace, "Parental Investment and the Optimization of Human Family Size," *Philosophical Transactions of the Royal Society B: Biological Sciences* 366, no. 1563 (2011): 333–43.

7. Fenton, Norman, "The Only Child," *The Pedagogical Seminary and Journal of Genetic Psychology* 35, no. 4 (1928): 546–56.

8. Juhn, Chinhui, Yona Rubinstein, and C. Andrew Zuppann, *The Quantity-Quality Trade-Off and the Formation of Cognitive and Non-Cognitive Skills*, National Bureau of Economic Research working paper no. 21824, December 2015.

9. Kristensen, Petter, and Tor Bjerkedal, "Explaining the Relation Between Birth Order and Intelligence," *Science* 316, no. 5832 (2007): 1717.

10. Koskela, Esa, "Offspring Growth, Survival and Reproductive Success in the Bank Vole: A Litter Size Manipulation Experiment," *Oecologia* 115, no. 3 (1998): 379–84.

11. Koivula, Minna, et al., "Cost of Reproduction in the Wild: Manipulation of Reproductive Effort in the Bank Vole," *Ecology* 84, no. 2 (2003): 398–405.

12. Sugimoto, Chikatoshi, et al., "Observations of Schooling Behaviour in the Oval Squid *Sepioteuthis lessoniana* in Coastal Waters of Okinawa Island," *Marine Biodiversity Records* 6 (2013).

13. Bieber, C., "Population Dynamics, Sexual Activity, and Reproduction Failure in the Fat Dormouse (Myoxus glis)," *Journal of Zoology* 244, no. 02 (1998): 223–29.

14. Gillespie, Duncan OS, Andrew F. Russell, and Virpi Lummaa, "When Fecundity Does Not Equal Fitness: Evidence of an Offspring Quantity Versus Quality Trade-Off in Pre-Industrial Humans," *Proceedings of the Royal Society of London B: Biological Sciences* 275, no. 1635 (2008): 713–22.

15. Skirbekk, Vegard, "Fertility Trends by Social Status," *Demographic Research* 18, no. 5 (2008): 145–80.

16. Rytkönen, Seppo, and Markku Orell, "Great Tits, Parus major, Lay Too Many Eggs: Experimental Evidence in Mid-Boreal Habitats," *Oikos* 93, no. 3 (2001): 439–50.

17. Bustamante, Javier, José J. Cuervo, and Juan Moreno, "The Function of Feeding Chases in the Chinstrap Penguin, *Pygoscelis Antarctica*," *Animal Behaviour* 44, no. 4 (1992): 753–59.

18. Jayachandran, Seema, and Ilyana Kuziemko, "Why Do Mothers Breastfeed Girls Less than Boys? Evidence and Implications for Child Health in India," *The Quarterly Journal of Economics* 126, no. 3 (2011): 1485–538.

19. Weimerskirch, Henri, Christophe Barbraud, and Patrice Lys, "Sex Differences in Parental Investment and Chick Growth in Wandering Albatrosses: Fitness Consequences," *Ecology* 81, no. 2 (2000): 309–18.

20. Trivers, Robert L., and Dan E. Willard, "Natural Selection of Parental Ability to Vary the Sex Ratio of Offspring," *Science* 179, no. 4068 (1973): 90–92.

21. Sunnucks, Paul, and Andrea C. Taylor, "Sex of Pouch Young Related to Maternal Weight in *Macropus eugenii* and *M. parma* (Marsupialia: Macropodidae)," *Australian Journal of Zoology* 45, no. 6 (1997): 573–78.

22. Schwartz, Christine R., "Earnings Inequality and the Changing Association Between Spouses' Earnings," *American Journal of Sociology* 115, no. 5 (2010): 1524.

23. Hopcroft, Rosemary L., and David O. Martin, "The Primary Parental Investment in Children in the Contemporary USA Is Education," *Human Nature* 25, no. 2 (2014): 235–50.

24. Almeida, David M., Elaine Wethington, and Amy L. Chandler, "Daily Transmission of Tensions Between Marital Dyads and Parent-Child Dyads," *Journal of Marriage and the Family* (1999): 49–61.

25. Chapman, Demian D., et al., "The Behavioural and Genetic Mating System of the Sand Tiger Shark, *Carcharias taurus*, an Intrauterine Cannibal," *Biology letters* 9, no. 3 (2013): 20130003.

26. Anderson, David J., "The Role of Parents in Sibilicidal Brood Reduction of Two Booby Species," *The Auk* (1995): 860–69.

27. Pollet, Thomas V., and Ashley D. Hoben, "An Evolutionary Perspective on Siblings: Rivals and Resources," *The Oxford Handbook of Evolutionary Family Psychology* (2011): 128–48.

28. Hodge, Sarah J., T. P. Flower, and T. H. Clutton-Brock, "Offspring Competition and Helper Associations in Cooperative Meerkats," *Animal Behaviour* 74, no. 4 (2007): 957–64.

29. Trillmich, Fritz, and Jochen BW Wolf, "Parent-Offspring and Sibling Conflict in Galápagos Fur Seals and Sea Lions," *Behavioral Ecology and Sociobiology* 62, no. 3 (2008): 363–75.

30. McGuire, Shirley, et al., "Children's Perceptions of Sibling Conflict During Middle Childhood: Issues and Sibling (Dis)similarity," *Social Development* 9, no. 2 (2000): 173–90.

31. Tucker, Corinna Jenkins, and Kerry Kazura, "Parental Responses to School-Aged Children's Sibling Conflict," *Journal of Child and Family Studies* 22, no. 5 (2013): 737–45.

32. Ibid.

33. Roulin, Alexandren, Mathias Kölliker, and Heinz Richner, "Barn Owl (*Tyto alba*) Siblings Vocally Negotiate Resources," *Proceedings of the Royal Society of London B: Biological Sciences* 267, no. 1442 (2000): 459–63.

34. Caro, Timothy M., *Cheetahs of the Serengeti Plains: Group Living in an Asocial Species* (Chicago: University of Chicago Press, 1994).

35. Caro, Timothy M., and D. A. Collins, "Male Cheetah Social Organization and Territoriality," *Ethology* 74, no. 1 (1987): 52–64.

CHAPTER 7

1. Begg, Colleen Margaret, "Feeding Ecology and Social Organisation of Honey Badgers (*Mellivom capensis*) in the Southern Kalahari," PhD dissertation, University of Pretoria, 2001.

2. Ibid.

3. Slagsvold, Tore, and Karen L. Wiebe, "Learning the Ecological Niche," *Proceedings of the Royal Society of London B: Biological Sciences* 274, no. 1606 (2007): 19–23.

4. Carruth, Betty Ruth, et al., "Prevalence of Picky Eaters Among Infants and Toddlers and Their Caregivers' Decisions About Offering a New Food," *Journal of the American Dietetic Association* 104 (2004): 57–64.

5. Birch, Leann L., "Research in Review. Children's Eating: The Development of Food-Acceptance Patterns," *Young Children* 50, no. 2 (1995): 71–78.

6. Skinner, Jean D., et al., "Do Food-Related Experiences in the First 2 Years of Life Predict Dietary Variety in School-Aged Children?" *Journal of Nutrition Education and Behavior* 34, no. 6 (2002): 310–15.

7. Di Bitetti, Mario S., et al., "Sleeping Site Preferences in Tufted Capuchin Monkeys (*Cebus apella nigritus*)," *American Journal of Primatology* 50, no. 4 (2000): 257–74.

8. Berk, Laura E., Trisha D. Mann, and Amy T. Ogan, "Make-Believe Play: Wellspring for Development of Self-Regulation," in *Play = Learning: How Play Motivates and Enhances Children's Cognitive and Social-Emotional Growth*, eds. Dorothy G. Singer, Roberta Michnick Golinkoff, and Kathy Hirsh-Pasek (New York: Oxford University Press, 2006), 74–100.

9. Mariette, Mylene M., and Katherine L. Buchanan, "Prenatal Acoustic Communication Programs Offspring for High Posthatching Temperatures in a Songbird," *Science* 353, no. 6301 (2016): 812–14.

10. Hamann, Katharina, et al., "Collaboration Encourages Equal Sharing in Children But Not in Chimpanzees," *Nature* 476, no. 7360 (2011): 328–31.

11. Bartal, Inbal Ben-Ami, Jean Decety, and Peggy Mason, "Empathy and Pro-Social Behavior in Rats," *Science* 334, no. 6061 (2011): 1427–30.

12. Sato, Nobuya, et al., "Rats Demonstrate Helping Behavior Toward a Soaked Conspecific," *Animal Cognition* 18, no. 5 (2015): 1039–47.

13. Bartal, Inbal Ben-Ami, et al., "Pro-Social Behavior in Rats Is Modulated by Social Experience," *Elife* 3 (2014): e01385.

14. Kestenbaum, Roberta, Ellen A. Farber, and L. Alan Sroufe, "Individual Differences in Empathy Among Preschoolers: Relation to Attachment History," *New Directions for Child and Adolescent Development* 1989, no. 44 (1989): 51–64.

15. Volk, Anthony A., et al., "Is Adolescent Bullying an Evolutionary Adaptation?" *Aggressive Behavior* 38, no. 3 (2012): 222–38.

16. Gallup, Andrew C., Daniel T. O'Brien, and David Sloan Wilson, "Intrasexual Peer Aggression and Dating Behavior During Adolescence: An Evolutionary Perspective," *Aggressive Behavior* 37, no. 3 (2011): 258–67.

17. Haig, David, "Transfers and Transitions: Parent-Offspring Conflict, Genomic Imprinting, and the Evolution of Human Life History," *Proceedings of the National Academy of Sciences* 107, no. suppl 1 (2010): 1731–35.

18. Margraf, Nicolas, and Andrew Cockburn, "Helping Behaviour and Parental Care in Fairy-Wrens (Malurus)," *Emu* 113, no. 3 (2013): 294–301.

19. Wahlström, L. K., and O. Liberg, "Patterns of Dispersal and Seasonal Migration in Roe Deer (*Capreolus capreolus*)," *Journal of Zoology* 235, no. 3 (1995): 455–67.

20. Clemens, Audra W., and Leland J. Axelson, "The Not-So-Empty-Nest: The Return of the Fledgling Adult," *Family Relations* (1985): 259–64.

21. Vespa, Jonathan, Jamie M. Lewis, and Rose M. Kreider, "America's Families and Living Arrangements: 2012," *Current Population Reports* 20 (2013): P570.

CHAPTER 8

1. Wong, Marian YL, et al., "The Threat of Punishment Enforces Peaceful Cooperation and Stabilizes Queues in a Coral-Reef Fish," *Proceedings of the Royal Society of London B: Biological Sciences* 274, no. 1613 (2007): 1093–99.

2. Lonsdorf, Elizabeth V., "What Is the Role of Mothers in the Acquisition of Termite-Fishing Behaviors in Wild Chimpanzees (*Pan troglodytes schweinfurthii*)?" *Animal Cognition* 9, no. 1 (2006): 36–46.

3. Gzesh, Steven M., and Colleen F. Surber, "Visual Perspective-Taking Skills in Children," *Child Development* (1985): 1204–13.

4. Fuster, Joaquin M., "Prefrontal Cortex," in *Comparative Neuroscience and Neurobiology*, ed. Louis N. Irwin (New York: Springer, 1988), 107–9.

5. Semple, Stuart, Melissa S. Gerald, and Dianne N. Suggs, "Bystanders Affect the Outcome of Mother-Infant Interactions in Rhesus Macaques," *Proceedings of the Royal Society of London B: Biological Sciences* 276, no. 1665 (2009): 2257–62.

6. Hart, Heledd, and Katya Rubia, "Neuroimaging of Child Abuse: A Critical Review," *Frontiers in Human Neuroscience* 6 (2012): 52.

7. Leonardi, Rebecca J., Sarah-Jane Vick, and Valérie Dufour, "Waiting for More: The Performance of Domestic Dogs (*Canis familiaris*) on Exchange Tasks," *Animal Cognition* 15, no. 1 (2012): 107–20.

8. Grosch, James, and Allen Neuringer, "Self-Control in Pigeons Under the Mischel Paradigm," *Journal of the Experimental Analysis of Behavior* 35, no. 1 (1981): 3–21.

9. Russell, Beth S., Rucha Londhe, and Preston A. Britner, "Parental Contributions to the Delay of Gratification in Preschool-Aged Children," *Journal of Child and Family Studies* 22, no. 4 (2013): 471–78.

10. Potegal, Michael, Michael R. Kosorok, and Richard J. Davidson, "Temper Tantrums in Young Children: 2. Tantrum Duration and Temporal Organization," *Journal of Developmental and Behavioral Pediatrics* 24, no. 3 (2003): 148–54.

11. Chang, Rosemarie Sokol, and Nicholas S. Thompson, "The Attention-Getting Capacity of Whines and Child-Directed Speech," *Evolutionary Psychology* 8, no. 2 (2010): 260–74.

12. Theunissen, Meinou HC, Anton GC Vogels, and Sijmen A. Reijneveld, "Punishment and Reward in Parental Discipline for Children Aged 5 to 6 Years: Prevalence and Groups at Risk," *Academic Pediatrics* 15, no. 1 (2015): 96–102.

13. Hart, Donna, and Robert W. Sussman, *Man the Hunted: Primates, Predators, and Human Evolution* (New York: Westview Press, 2005).

14. National Center for Missing and Exploited Children, "Key Facts," missingkids .org/KeyFacts, accessed January 15, 2017.

15. Estes and Estes, "The Birth and Survival of Wildebeest Calves."

16. Weinpress, Meghan, "Maternal and Alloparental Discipline in Atlantic Spotted Dolphins (*Stenella frontalis*) in the Bahamas," (master's thesis, Florida Atlantic University, 2013).

17. Thornton, Alex, and Katherine McAuliffe, "Teaching in Wild Meerkats," *Science* 313, no. 5784 (2006): 227–29.

18. Foster, Emma A., et al., "Adaptive Prolonged Post-Reproductive Life Span in Killer Whales," *Science* 337, no. 6100 (2012): 1313.

19. Schiffrin, Holly H., et al., "Helping or Hovering? The Effects of Helicopter Parenting on College Students' Well-Being," *Journal of Child and Family Studies* 23, no. 3 (2014): 548–57.

20. Mineka, Susan, et al., "Observational Conditioning of Snake Fear in Rhesus Monkeys," *Journal of Abnormal Psychology* 93, no. 4 (1984): 355.

21. Cook, Michael, and Susan Mineka, "Observational Conditioning of Fear to Fear-Relevant Versus Fear-Irrelevant Stimuli in Rhesus Monkeys," *Journal of Abnormal Psychology* 98, no. 4 (1989): 448.

22. Etting, Stephanie F., Lynne A. Isbell, and Mark N. Grote, "Factors Increasing Snake Detection and Perceived Threat in Captive Rhesus Macaques (*Macaca mulatta*)," *American Journal of Primatology* 76, no. 2 (2014): 135–45.

23. DeLoache, Judy S., and Vanessa LoBue, "The Narrow Fellow in the Grass: Human Infants Associate Snakes and Fear," *Developmental Science* 12, no. 1 (2009): 201–7.

24. LoBue, Vanessa, et al., "Young Children's Interest in Live Animals," *British Journal of Developmental Psychology* 31, no. 1 (2013): 57–69.

25. Thrasher, Cat, and Vanessa LoBue, "Do Infants Find Snakes Aversive? Infants' Physiological Responses to 'Fear-Relevant' Stimuli," *Journal of Experimental Child Psychology* 142 (2016): 382–90.

26. Debiec, Jacek, and Regina Marie Sullivan, "Intergenerational Transmission of Emotional Trauma Through Amygdala-Dependent Mother-to-Infant Transfer of Specific Fear," *Proceedings of the National Academy of Sciences* 111, no. 33 (2014): 12222–27.

27. LeMoyne, Terri, and Tom Buchanan, "Does 'Hovering" Matter? Helicopter Parenting and Its Effect on Well-Being," *Sociological Spectrum* 31, no. 4 (2011): 399–418.

28. Schiffrin, et al., "Helping or Hovering? The Effects of Helicopter Parenting on College Students' Well-Being."

29. US Department of Health and Human Services, Administration for Children and Families, Administration on Children, Youth and Families, Children's Bureau, *Child Maltreatment 2009* and *Child Maltreatment 2010*, available at acf.hhs.gov/programs/cb/stats_research/index.htm#can, accessed January 15, 2017.

30. US Department of Health and Human Services, Administration for Children and Families, Administration on Children, Youth and Families, Children's Bureau, *Child Maltreatment 2015*, acf.hhs.gov/sites/default/files/cb/cm2015.pdf#page=20, accessed January 28, 2017.

31. Bustnes, Jan O., Kjell E. Erikstad, and Tor H. Bjørn, "Body Condition and Brood Abandonment in Common Eiders Breeding in the High Arctic," *Waterbirds* 25, no. 1 (2002): 63–66.

32. Bustnes, Jan O., and Kjell E. Erikstad, "Parental Care in the Common Eider (*Somateria mollissima*): Factors Affecting Abandonment and Adoption of Young," *Canadian Journal of Zoology* 69, no. 6 (1991): 1538–45.

33. Begle, Angela Moreland, Jean E. Dumas, and Rochelle F. Hanson, "Predicting Child Abuse Potential: An Empirical Investigation of Two Theoretical Frameworks," *Journal of Clinical Child and Adolescent Psychology* 39, no. 2 (2010): 208–19.

34. Maestripieri, Dario, and Kelly A. Carroll, "Child Abuse and Neglect: Usefulness of the Animal Data," *Psychological Bulletin* 123, no. 3 (1998): 211.

35. Simons, Dominique A., and Sandy K. Wurtele, "Relationships Between Parents' Use of Corporal Punishment and Their Children's Endorsement of Spanking and Hitting Other Children," *Child Abuse and Neglect* 34, no. 9 (2010): 639–46.

36. Tomoda, Akemi, et al., "Reduced Prefrontal Cortical Gray Matter Volume in Young Adults Exposed to Harsh Corporal Punishment," *Neuroimage* 47 (2009): T66–T71.

37. Committee on the Rights of the Child, General Comment No. 8, "The Right of the Child to Protection from Corporal Punishment and Other Cruel or Degrading Forms of Punishment," ohchr.org/EN/HRBodies/CRC/Pages/CRCIndex.aspx, accessed January 15, 2017.

38. Glaser, Danya, "Emotional Abuse and Neglect (Psychological Maltreatment): A Conceptual Framework," *Child Abuse and Neglect* 26, no. 6 (2002): 697–714.

CHAPTER 9

1. Brown, Jason L., Victor Morales, and Kyle Summers, "A Key Ecological Trait Drove the Evolution of Biparental Care and Monogamy in an Amphibian," *The American Naturalist* 175, no. 4 (2010): 436–46.

2. Goldstein, Joseph, Anna Freud, and Albert J. Solnit, *Beyond the Best Interests of the Child* (1980): 22.

3. Walper, Sabine, Carolin Thönnissen, and Philipp Alt, "Effects of Family Structure and the Experience of Parental Separation: A Study on Adolescents' Well-Being," *Comparative Population Studies* 40, no. 3 (2015).

4. Ahern, Todd H., Elizabeth AD Hammock, and Larry J. Young, "Parental Division of Labor, Coordination, and the Effects of Family Structure on Parenting in Monogamous Prairie Voles (*Microtus ochrogaster*)," *Developmental Psychobiology* 53, no. 2 (2011): 118–31.

5. Creighton, J. Curtis, et al., "Dynamics of Biparental Care in a Burying Beetle: Experimental Handicapping Results in Partner Compensation," *Behavioral Ecology and Sociobiology* 69, no. 2 (2015): 265–71.

6. Fetherston, Isabelle A., Michelle Pellissier Scott, and James FA Traniello, "Behavioural Compensation for Mate Loss in the Burying Beetle *Nicrophorus orbicollis*," *Animal Behaviour* 47, no. 4 (1994): 777–85.

7. Harrison, F., Z. Barta, I. Cuthill, and Tamas Szekely, "How Is Sexual Conflict over Parental Care Resolved? A Meta-Analysis," *Journal of Evolutionary Biology* 22, no. 9 (2009): 1800–1812.

8. Walper, Thönnissen, and Alt, "Effects of Family Structure and the Experience of Parental Separation: A Study on Adolescents' Well-Being."

9. Laurenson, M. Karen, "Cub Growth and Maternal Care in Cheetahs," *Behavioral Ecology* 6, no. 4 (1995): 405–9.

10. Walters, Jeffrey R., Phillip D. Doerr, and J. H. Carter, "The Cooperative Breeding System of the Red-Cockaded Woodpecker," *Ethology* 78, no. 4 (1988): 275–305.

11. Conner, Richard N., et al., "Group Size and Nest Success in Red-Cockaded Woodpeckers in the West Gulf Coastal Plain: Helpers Make a Difference," *Journal of Field Ornithology* 75, no. 1 (2004): 74–78.

12. White, Angela M., and Elissa Z. Cameron, "Evidence of Helping Behavior in a Free-Ranging Population of Communally Breeding Warthogs," *Journal of Ethology* 29, no. 3 (2011): 419–25.

13. Sarkisian, N., and N. Gerstel, "Does Singlehood Isolate or Integrate? Examining the Link Between Marital Status and Ties to Kin, Friends, and Neighbors," *Journal of Social and Personal Relationships* 33, no. 3 (2016): 361–84.

14. Fernández, Gustavo J., and Juan C. Reboreda, "Male Parental Care in Greater Rheas (*Rhea americana*) in Argentina." *The Auk* 120, no. 2 (2003): 418–28.

15. Forsgren, Elisabet, Anna Karlsson, and Charlotta Kvarnemo, "Female Sand Gobies Gain Direct Benefits by Choosing Males with Eggs in Their Nests," *Behavioral Ecology and Sociobiology* 39, no. 2 (1996): 91–96.

16. Bygott, J. David, Brian CR Bertram, and Jeannette P. Hanby, "Male Lions in Large Coalitions Gain Reproductive Advantages," *Nature* 282 (1979): 839–41.

17. Veiga, José P., "Infanticide by Male and Female House Sparrows," *Animal Behaviour* 39, no. 3 (1990): 496–502.

18. Veiga, José P., "Replacement Female House Sparrows Regularly Commit Infanticide: Gaining Time or Signaling Status?" *Behavioral Ecology* 15, no. 2 (2004): 219–22.

19. Tooley, Greg A., et al., "Generalising the Cinderella Effect to Unintentional Childhood Fatalities." *Evolution and Human Behavior* 27, no. 3 (2006): 224–30.

20. Anderson, Kermyt G., Hillard Kaplan, and Jane Lancaster, "Paternal Care by Genetic Fathers and Stepfathers I: Reports from Albuquerque Men," *Evolution and Human Behavior* 20, no. 6 (1999): 405–31.

21. Hector, Anne C. Keddy, Robert M. Seyfarth, and Micheal J. Raleigh, "Male Parental Care, Female Choice and the Effect of an Audience in Vervet Monkeys," *Animal Behaviour* 38, no. 2 (1989): 262–71.

22. Meek, Susan B., and Raleigh J. Robertson, "Adoption of Young by Replacement Male Birds: An Experimental Study of Eastern Bluebirds and a Review," *Animal Behaviour* 42, no. 5 (1991): 813–20.

23. Henry, Katherine A., and Mienah Zulfacar Sharif, "Historical and Policy Perspectives of Child Health in the United States," *Child Health: A Population Perspective* (2015): 9.

24. Dunham, Noah Thomas, and Paul Otieno Opere, "A Unique Case of Extra-Group Infant Adoption in Free-Ranging Angola Black and White Colobus Monkeys (*Colobus angolensis palliatus*)," *Primates* 57, no. 2 (2016): 187–94.

25. Brodzinsky, D., and Ellen Pinderhughes, "Parenting and Child Development in Adoptive Families," *Handbook of parenting* 1 (2013): 279–311.

26. Woolfenden, Glen Everett, and John W. Fitzpatrick, *The Florida Scrub Jay: Demography of a Cooperative-Breeding Bird*, Vol. 20. (Princeton, NJ: Princeton University Press, 1984).

27. Young, Lindsay C., and Eric A. VanderWerf, "Adaptive Value of Same-Sex Pairing in Laysan Albatross," *Proceedings of the Royal Society of London B: Biological Sciences* 281, no. 1775 (2014): 20132473.

28. Clutton-Brock, T. H., A. F. Russell, and L. L. Sharpe, "Meerkat Helpers Do Not Specialize in Particular Activities," *Animal Behaviour* 66, no. 3 (2003): 531–40.

29. Du Plessis, Morné A., "Helping Behaviour in Cooperatively-Breeding Green Woodhoopoes: Selected or Unselected Trait?" *Behaviour* 127, no. 1 (1993): 49–65.

30. Hrdy, Sarah Blaffer, *Mothers and Others: The Evolutionary Origins of Mutual Understanding* (Cambridge, MA: Harvard University Press, 2009).

31. Ibid.

ACKNOWLEDGMENTS

I want to thank the many people who made this book a reality. First and foremost is my agent extraordinaire, Uwe, for believing that approaches that "fall between the stools" have a place. Thank you for always encouraging me. So many individuals (human and other animals) inspired me on this journey. From those early childhood friends and surrogate families all the way through my adult surrogate family. Thank you Mary Barerra for all the adventures with many more to come, Mrs. B for giving me a safe haven as a child and loving me just the way I am, and Patti Ragan, Director of the Center for Great Apes, for inspiring me, embracing me when I was an adrift young woman, and continuing to welcome me "home." A heartfelt and grateful thank you to Ramona Walls for taking me in and welcoming me into your family, Melissa Mark for being the sister I wish I had, and Christopher Jensen for encouraging me to let my light shine. Equally important were all the people who contributed intellectually to this book by reviewing and critiquing chapters, discussing concepts and ideas, and listening to me go on (and on) about the latest interesting find I uncovered: Christopher Jensen, Ramona Walls, Melissa Mark, Rita Mitra, Dominique Boykin, and many more. These lively discussions, perspectives, and generous contributions of ideas enhanced this book immeasurably. I want to give a special thank you to Nicholas Cizek, my editor at The Experiment, for "getting me," and, through outstanding editing, enriching the many messages this book conveys. And of course, thank you to the entire Experiment team for all the effort and work that has gone into making this book a reality.

ABOUT THE AUTHOR

DR. JENNIFER L. VERDOLIN is an adjunct professor in the Department of Biology at Duke University. She received her BS degree from Florida Atlantic University and her MS from Northern Arizona University, and earned her PhD in ecology and evolution at Stony Brook University in 2008. Verdolin's scientific research explores the evolution of social behavior and mating systems. Her first popular science book, *Wild Connection: What Animal Courtship and Mating Tell Us About Human Relationships*, was released in 2014. She is the featured guest of the segment "Think Like a Human, Act Like an Animal" on the nationally syndicated *D.L. Hughley Show*, and was featured in the BBC One Documentary *Animals in Love*.